Qualitative Analysis and the Properties of Ions in Aqueous Solution

EMIL J. SLOWINSKI

Professor of Chemistry
Macalester College
St. Paul, Minnesota

WILLIAM L. MASTERTON

Professor of Chemistry
University of Connecticut
Storrs, Connecticut

1971
W. B. SAUNDERS COMPANY
PHILADELPHIA · LONDON · TORONTO

 Saunders Golden Series

W. B. Saunders Company: West Washington Square
Philadelphia, Pa. 19105

12 Dyott Street
London, WC1A 1DB

1835 Yonge Street
Toronto 7, Canada

Qualitative Analysis and the Properties of Ions   SBN   0-7216-8369-X
in Aqueous Solution

Print No.:   9   8   7   6   5   4   3   2   1

PREFACE

One of the most common complaints of the more practical minded chemists, and chemistry teachers, is that modern courses in the undergraduate chemistry curriculum, in particular the introductory course, do not contain enough descriptive chemistry. In our general chemistry text, *Chemical Principles,* we decided to omit some of the detailed descriptive chemistry found in many other introductory texts, not because we felt that such descriptive chemistry was unimportant, but because, in our opinion at least, descriptive chemistry is most readily learned in the laboratory, where the chemicals are actually used in carrying out chemical reactions.

Laboratory work on qualitative analysis of cations and anions is one way of training a beginning student in inorganic descriptive chemistry. Students are exposed to a great many properties of the ions, observe many different compounds, which they prepare by many different reactions, and learn to write equations for these reactions. The use of unknowns adds an element of interest that makes the study much more pleasant and introduces some of the atmosphere that is present in chemical research. Although the development of modern instrumental methods has greatly reduced the practical utility of classical qualitative analysis, it remains a very valuable teaching tool. We believe it will continue to be a popular method for integrating descriptive chemistry and theory in elementary courses.

Several qualitative analysis texts are in current use, and one might wonder why we write still another. Having used such texts in various ways over the years, we feel that their usual drawback is that they over-emphasize the properties of an ion that are actually the basis of the standard qualitative analysis scheme. This limitation seems most unfortunate in that the student may never hear about some properties of an ion that are at least as important as those necessary for analysis. More important, he becomes tied to a standard procedure for analysis and may not realize that, under some circumstances, he can develop his own very effective schemes of analysis.

In this text, in addition to a detailed scheme for the qualitative analysis of the common cations and anions, we include a short discussion of the chemical properties of each of the ions. For each group of ions the first laboratory assignment requires that the student become familiar with the standard procedure for analysis of that group and use the procedure to find the composition of an unknown. If the student examines the individual prop-

erties of the ions in the group, he can see that the separation and identification
of the ions result from a rational application of chemical principles used to
take advantage of the properties of individual ions.

Perhaps the most unusual feature of this text is the laboratory assignments.
We depart from the traditional procedure of requiring students to analyze
group unknowns that may contain any of the ions in the particular group.
Instead, we suggest assigning unknowns consisting of a limited number of ions
chosen from two or more groups. For example, a student might be asked to
analyze an unknown containing no cations other than Ag^+, Cu^{2+}, and Fe^{3+}.
He is expected to use his knowledge of the properties of these ions to develop
an abbreviated scheme of analysis, omitting many of the steps in the general
procedure for Group I, II, and III cations. Hopefully, he will go one step
further and devise a simple series of tests entirely independent of the general
analytical scheme. For example, he might add 6M NH_3 to precipitate
$Fe(OH)_3$ and leave the other two ions in solution as the ammine complexes
$Ag(NH_3)_2{}^+$ and $Cu(NH_3)_4{}^{2+}$. Having identified $Fe(OH)_3$ and $Cu(NH_3)_4{}^{2+}$ by
their characteristic colors, the student could then test for Ag^+ by acidifying
the solution with 6M HCl to obtain a precipitate of AgCl.

This example is illustrative of the types of laboratory assignments we
include to stimulate the student to set up his own schemes of analysis. To
do this successfully he must know the general principles of chemical equilibrium and how to apply them in qualitative analysis, and he must also be
familiar with the properties of the ions involved. We feel that the average
student, given appropriate guidance and carefully selected problems, can
make good progress in this kind of analysis. It has the advantage that it requires
the student to think as a chemist rather than to rely on methods furnished
to him by others. Superimposed on a study of the conventional procedures,
the development of schemes of analysis by the student should add a worthwhile
dimension to the usual course in qualitative analysis.

Perhaps a word about terminology and notation would be appropriate.
Since we feel that students should become acquainted with the common
systems, we occasionally use each of them. Therefore, one finds reference to
cuprous and cupric ions as well as to copper(I) and copper(II) species. Unless
necessary to an explanation, we ignore the fact that many cations are hydrated
or otherwise complexed, and write, for example, Cu^{2+}, Cr^{3+}, Bi^{3+}, Sn^{4+}, and,
rarely, even Sb^{5+}. We realize that the latter two ions are probably always
present in solution as complex anions, but feel that if we wrote these species
as Sn(IV) and Sb(V), we probably should also consider Bi^{+3} and Hg^{2+} as special
cases, since bismuth in solution may also be a complex anion and mercury(II)
compounds often show very little ionic character. After much agonizing
indecision on this matter, we decided on the notation we indicate, admittedly
in part because all these species are included in the procedures for analysis
of cations; on that basis, it seemed only fair to admit the possibility that they
could be described, at least formally, as cations.

The procedures we use in our standard scheme of analysis are similar to those in most texts on the subject. In cases in which the classic procedures did not work well when we applied them, we have made changes. Some of the discussions are based on similar material in *Qualitative Chemical Analysis,* by L. J. Curtman, published by Macmillan and now out of print. We would like to acknowledge the helpful assistance of Mr. Daniel White, who spent a summer at Macalester testing all the standard procedures from the student's point of view. Although we have done our best to use effective procedures, it is possible that there are a few difficulties remaining hidden in them. We would very much appreciate any comments, criticisms, or suggestions from chemistry teachers and students who use this book.

EMIL J. SLOWINSKI
WILLIAM L. MASTERTON

TABLE OF CONTENTS

CHAPTER 1

THE THEORY AND PRACTICE OF QUALITATIVE ANALYSIS

One of the first things that a chemist must find out about a sample of matter with which he is working is the nature of the components. The determination of the species present in a sample is in the area of chemistry known as qualitative analysis. Once we know what substances are in a sample we may go further and carry out a quantitative analysis to establish how much of each substance is there. In this course we will restrict our attention to problems of qualitative analysis.

Obviously, the difficulty of an analysis depends tremendously on the nature of the sample. The qualitative analysis of the fluid present in brain cells or of the atmosphere over a large industrial city may require the most sophisticated techniques available, whereas the determination of the major components of sea water might be a relatively easy matter. The problems with which we will deal here are for the most part in the simple, but not trivial, category. You will be asked to identify the cations, or anions, or both, in various "unknown" solutions. The schemes of analysis you use will often be furnished to you; however, sometimes you will be asked to set up your own procedures, since there is nothing magic or especially difficult about the methods to be employed.

The main portion of the text will be devoted to discussion of the properties of some rather common cations and anions and the procedures used to establish the presence of these species in aqueous solution. The ions we will consider are the following:

Cations: Ag^+, Pb^{2+}, Hg_2^{2+}, Bi^{3+}, Cu^{2+}, Cd^{2+}, Hg^{2+}, Sn^{2+} or Sn^{4+}, Sb^{3+} or Sb^{5+}, Fe^{2+} or Fe^{3+}, Al^{3+}, Mn^{2+}, Zn^{2+}, Ni^{2+}, Cr^{3+}, Co^{2+}, Ba^{2+}, Ca^{2+}, Mg^{2+}, Na^+, K^+, and NH_4^+

Anions: CO_3^{2-}, SO_4^{2-}, PO_4^{3-}, SO_3^{2-}, $C_2O_4^{2-}$, CrO_4^{2-}, Cl^-, Br^-, I^-, SCN^-, S^{2-}, NO_3^-, NO_2^-, ClO_3^-, and $C_2H_3O_2^-$

Although the list is rather long, including 25 cations and 15 anions, it is by no means exhaustive. We have omitted the more uncommon species and have intentionally avoided ions that are particularly hazardous or difficult to work with.

THE GENERAL METHOD IN QUALITATIVE ANALYSIS

The development of a general procedure to analyze a solution that might contain any of the cations or anions on our list would indeed be a formidable problem. One of your tasks in this course is to become reasonably well acquainted with one such procedure. You will find it rational and relatively easy to understand, but, in all probability, not intuitively obvious. The general approach to the problem of qualitative analysis, however, is rather obvious and worth considering. We will now illustrate the approach in a simple case; the principles we will use still apply when the system is much more complex.

Let us assume that we are interested in cation analysis, and that we have a sample that we are told might contain Pb^{2+}, Cd^{2+}, and Zn^{2+}, but no other cations. Our problem is to set up a systematic procedure to allow us to establish the presence or absence of those cations in a solution of "unknown" composition.

After perhaps a false start or two, we would probably obtain solutions containing only one of the cations and perform experiments on those "known" solutions to find out some of the chemical properties of the cations. Our purpose would be to look for differences in the properties of the cations, with a view toward using those differences to identify the ions. One of the simplest things we might do is add solutions of some common reagents to a salt solution containing one of the cations and note any changes that might indicate that a reaction is occurring. With some reagents, we might get a solid precipitate, a change in color, or the evolution of a gas.

A very common reaction with reagents such as 6M NH_3, 6M NaOH, 6M H_2SO_4, and 1M Na_2CO_3 is precipitation. The cations in our example differ remarkably in their solubility behavior toward these reagents; in some cases there is no precipitate and no apparent reaction, but in others a precipitate forms, which may or may not be soluble in an excess of that reagent. Our observations on testing each cation solution with each reagent are summarized in Table 1-1. In the table, P means that a precipitate forms on addition of the reagent; D, that the precipitate dissolves in excess reagent; and S, that no precipitate forms. (Incidentally, all the solutions are colorless and all the precipitates are white.)

After a little cogitation, and just possibly a little experience, one would

TABLE 1-1. **Effects of Reagents on Cation Solutions**

	NH_3	$NaOH$	H_2SO_4	Na_2CO_3
Pb^{2+}	P	P, D	P	P
Cd^{2+}	P, D	P	S	P
Zn^{2+}	P, D	P, D	S	P

conclude that it would be rather easy to analyze a solution for the three cations, given the information in the table. There are probably several approaches one might use, but we will consider only one of them. As a first step, let us add 6M NaOH in excess to the mixture. If a precipitate remains, cadmium ion must be present, because the other two cations are soluble under such conditions; if there is no precipitate, cadmium is absent. So far, so good.

Since the cadmium, if present, is now in the solid precipitate, it can be separated from the solution, which now can contain only lead and zinc ions. The separation can be easily accomplished by filtration or centrifugation. If H_2SO_4 is then added to the solution in excess, lead should precipitate but zinc should remain in solution; a precipitate forming at this point means that Pb^{2+} was in the original solution. If no precipitate forms, no lead was there.

We then separate the solid from the remaining solution, which now can contain only Zn^{2+} from the original group. The presence of zinc would be established if addition of 1M Na_2CO_3 caused a precipitate to form.

The procedure for the analysis was developed on the basis of the properties of the three cations and really arises very logically. Indeed, you may be convinced at this stage that the approach would work perfectly. Unfortunately, if you test the procedure with a solution known to contain Pb^{2+}, Cd^{2+}, and Zn^{2+}, you may easily not get positive evidence of zinc ion, even though it is present. On adding Na_2CO_3, you would observe, perhaps to your surprise, bubbles of gas being evolved. If you persist in adding Na_2CO_3 until the bubbles stop forming, a precipitate does form. However, as long as gas is being given off, the solution remains clear. It should be apparent that something in the system is different from the situation indicated by the table, since there we observed no bubbles at any time.

The problem we have cited is illustrative in many ways of the situation in general qualitative analysis. We base the general scheme on the many differences in properties of the ions with respect to a rather small group of common reagents. In order for the scheme to work, however, we not only need to have information analogous to that given in the table, but we also need to know in detail what actually happens in a reaction, what species are formed, and how those species might be expected to behave under changing conditions.

In order, then, to develop and apply a scheme of qualitative analysis, we need to be able to work with two kinds of information. We must have the experimental facts—the empirical properties of the species with which we are working—similar to the information in the table. We must also know a good deal about the theoretical principles that govern the chemical reactions that occur in aqueous solutions. These reactions are of several kinds and include precipitations, acid-base reactions, complex ion formation, and oxidation-reduction reactions. Each of these reactions is subject to the general law of chemical equilibrium and can be controlled by taking advantage of the conditions imposed on the reaction by the law.

In the rest of this chapter we will consider the law of chemical equilibrium and its relationship to the various kinds of reactions we have just mentioned. Since our presentation here must be very brief, consisting essentially of only a review of the relevant principles, it would be highly advisable for you to reread the material in these areas in your general chemistry text, as you proceed further in this chapter.

THE LAW OF CHEMICAL EQUILIBRIUM

When a chemical reaction occurs, the reaction frequently appears to stop at a point at which some of each reactant is still present, rather than proceeding to completion. At that point one can consider that the rate at which the product is forming is just equal to the rate at which the product is reacting to reform the initial reactants. Such a system is said to be in a state of chemical equilibrium.

A system in chemical equilibrium is subject to a condition that governs the concentrations of the reactants and products and that can be expressed in a simple mathematical form. For a typical chemical reaction occurring in solution:

$$aA(\text{solute}) + bB(\text{solute}) \rightleftharpoons cC(\text{solid}) + dD(\text{solute})$$

the reaction proceeds until the following condition is satisfied:

$$K = \frac{[D]^d}{[A]^a[B]^b}$$

where K is a number, called the *equilibrium constant* for the reaction, [A], [B], and [D] are the equilibrium concentrations in moles per liter of the species A, B, and D, respectively, and a, b, and d are the coefficients in the balanced equation. Pure solid C does not appear in the expression for K but must be present in at least a small amount in the equilibrium system. For every chemical reaction, K has a characteristic value, which does not depend on the manner in which the equilibrium system was set up and which remains constant as long as the temperature does not change.

In the general case the equilibrium expression looks rather abstract, but it does permit one to draw a few important conclusions. If the value of K for a reaction is large, the equilibrium system will tend to have high concentrations of products relative to those of reactants. A small value of K implies that the reaction to form products will not proceed to any great extent.

From the form of the expression for K, it is clear that the equilibrium constant for the reverse reaction is always related to that for the forward reaction by the equation,

$$K_{\text{reverse}} = 1/K_{\text{forward}}$$

For some reactions, it is convenient, or traditional, to write the equation for the reaction in a particular direction; the equilibrium state of the system is not affected by doing this, but the mathematical treatment may be much facilitated.

Some reactions can be expressed as the sum of two or more somewhat more basic reactions. For example, consider

I $aA(s) \rightleftharpoons bB(solute) + cC(solute)$ $K_I = [B]^b[C]^c$

II $bB(solute) + dD(solute) \rightleftharpoons eE(solute)$ $K_{II} = [E]^e/([B]^b[D]^d)$

III $aA(s) + dD(solute) \rightleftharpoons cC(solute) + eE(solute)$ $K_{III} = [C]^c[E]^e/[D]^d$

Reaction III is the sum of reactions I and II. Since all three equilibria must obtain in the system if it is in a state of equilibrium, it must be true, as implied by the expressions for the K's, that

$$K_{III} = K_I \times K_{II}$$

This relation between the K's for different reactions is a useful one and can be stated in a general way:

> In a system involving multiple equilibria, the value of the equilibrium constant for a reaction, which can be expressed as the *sum* of several reactions, is equal to the *product* of the equilibrium constants of those reactions.

The law of chemical equilibrium furnishes the chemist with one of his most potent tools for understanding and controlling chemical reactions. Since the use of this important concept is most apparent in specific cases, we shall now proceed to consider how it is applied to the different kinds of reactions that are encountered in qualitative analysis.

PRECIPITATION REACTIONS

When two solutions containing ions are mixed, one often observes the formation of a solid precipitate. In the illustration discussed earlier of the properties of Pb^{2+}, Cd^{2+}, and Zn^{2+}, we noted that mixing 6M NaOH with a solution of cadmium nitrate, for example, produces a precipitate. In principle, the precipitate might be $NaNO_3$ or $Cd(OH)_2$. However, since all sodium salts are soluble, we can conclude that the precipitate must be cadmium

TABLE 1-2.   **Solubility Products at 25° C**

Compound	Reaction	K_{sp}
Aluminium		
Hydroxide	$Al(OH)_3 \rightleftharpoons Al^{3+} + 3OH^-$	1×10^{-32}
Phosphate	$AlPO_4 \rightleftharpoons Al^{3+} + PO_4{}^{3-}$	6.3×10^{-19}
Barium		
Carbonate	$BaCO_3 \rightleftharpoons Ba^{2+} + CO_3{}^{2-}$	5×10^{-9}
Chromate	$BaCrO_4 \rightleftharpoons Ba^{2+} + CrO_4{}^{2-}$	1.2×10^{-10}
Oxalate	$BaC_2O_4 \rightleftharpoons Ba^{2+} + C_2O_4{}^{2-}$	1.5×10^{-8}
Sulfate	$BaSO_4 \rightleftharpoons Ba^{2+} + SO_4{}^{2-}$	1.0×10^{-10}
Bismuth		
Hydroxide	$Bi(OH)_3 \rightleftharpoons Bi^{3+} + 3OH^-$	4×10^{-31}
Phosphate	$BiPO_4 \rightleftharpoons Bi^{3+} + PO_4{}^{3-}$	1.2×10^{-23}
Sulfide	$Bi_2S_3 \rightleftharpoons 2Bi^{3+} + 3S^{2-}$	1×10^{-97}
Cadmium		
Hydroxide	$Cd(OH)_2 \rightleftharpoons Cd^{2+} + 2OH^-$	2×10^{-14}
Oxalate	$CdC_2O_4 \rightleftharpoons Cd^{2+} + C_2O_4{}^{2-}$	1.8×10^{-8}
Sulfide	$CdS \rightleftharpoons Cd^{2+} + S^{2-}$	1×10^{-26}
Calcium		
Carbonate	$CaCO_3 \rightleftharpoons Ca^{2+} + CO_3{}^{2-}$	4.8×10^{-9}
Hydroxide	$Ca(OH)_2 \rightleftharpoons Ca^{2+} + 2OH^-$	5.5×10^{-6}
Oxalate	$CaC_2O_4 \rightleftharpoons Ca^{2+} + C_2O_4{}^{2-}$	1.3×10^{-9}
Phosphate	$Ca_3(PO_4)_2 \rightleftharpoons 3Ca^{2+} + 2PO_4{}^{3-}$	1×10^{-29}
Sulfate	$CaSO_4 \rightleftharpoons Ca^{2+} + SO_4{}^{2-}$	1.0×10^{-5}
Chromium(III)		
Hydroxide	$Cr(OH)_3 \rightleftharpoons Cr^{3+} + 3OH^-$	6×10^{-31}
Phosphate	$CrPO_4 \rightleftharpoons Cr^{3+} + PO_4{}^{3-}$	2.4×10^{-23}
Cobalt(II)		
Hydroxide	$Co(OH)_2 \rightleftharpoons Co^{2+} + 2OH^-$	1.3×10^{-15}
Sulfide	$CoS \rightleftharpoons Co^{2+} + S^{2-}$	1×10^{-21}
Copper(II)		
Hydroxide	$Cu(OH)_2 \rightleftharpoons Cu^{2+} + 2OH^-$	2.2×10^{-20}
Oxalate	$CuC_2O_4 \rightleftharpoons Cu^{2+} + C_2O_4{}^{2-}$	2.9×10^{-8}
Sulfide	$CuS \rightleftharpoons Cu^{2+} + S^{2-}$	6×10^{-36}
Iron(II)		
Carbonate	$FeCO_3 \rightleftharpoons Fe^{2+} + CO_3{}^{2-}$	3.5×10^{-11}
Hydroxide	$Fe(OH)_2 \rightleftharpoons Fe^{2+} + 2OH^-$	2.2×10^{-15}
Sulfide	$FeS \rightleftharpoons Fe^{2+} + S^{2-}$	6×10^{-18}
Iron(III)		
Hydroxide	$Fe(OH)_3 \rightleftharpoons Fe^{3+} + 3OH^-$	2.5×10^{-39}
Phosphate	$FePO_4 \rightleftharpoons Fe^{3+} + PO_4{}^{3-}$	1×10^{-22}
Lead		
Carbonate	$PbCO_3 \rightleftharpoons Pb^{2+} + CO_3{}^{2-}$	1.0×10^{-13}
Chloride	$PbCl_2 \rightleftharpoons Pb^{2+} + 2Cl^-$	1.6×10^{-5}
Chromate	$PbCrO_4 \rightleftharpoons Pb^{2+} + CrO_4{}^{2-}$	1.6×10^{-14}
Hydroxide	$Pb(OH)_2 \rightleftharpoons Pb^{2+} + 2OH^-$	1.2×10^{-15}
Oxalate	$PbC_2O_4 \rightleftharpoons Pb^{2+} + C_2O_4{}^{2-}$	8.3×10^{-12}
Phosphate	$Pb_3(PO_4)_2 \rightleftharpoons 3Pb^{2+} + 2PO_4{}^{3-}$	1×10^{-42}
Sulfate	$PbSO_4 \rightleftharpoons Pb^{2+} + SO_4{}^{2-}$	1.6×10^{-8}
Sulfide	$PbS \rightleftharpoons Pb^{2+} + S^{2-}$	1×10^{-27}

TABLE 1-2. (*Continued*)

Compound	Reaction	K_{sp}
Magnesium		
Ammonium		
Phosphate	$MgNH_4PO_4 \rightleftharpoons Mg^{2+} + NH_4^+ + PO_4^{3-}$	2.5×10^{-13}
Carbonate	$MgCO_3 \rightleftharpoons Mg^{2+} + CO_3^{2-}$	1×10^{-5}
Hydroxide	$Mg(OH)_2 \rightleftharpoons Mg^{2+} + 2OH^-$	1.1×10^{-11}
Oxalate	$MgC_2O_4 \rightleftharpoons Mg^{2+} + C_2O_4^{2-}$	8.6×10^{-5}
Manganese		
Carbonate	$MnCO_3 \rightleftharpoons Mn^{2+} + CO_3^{2-}$	2.0×10^{-11}
Hydroxide	$Mn(OH)_2 \rightleftharpoons Mn^{2+} + 2OH^-$	1.7×10^{-13}
Oxalate	$MnC_2O_4 \rightleftharpoons Mn^{2+} + C_2O_4^{2-}$	1.1×10^{-15}
Sulfide	$MnS \rightleftharpoons Mn^{2+} + S^{2-}$	7.1×10^{-16}
Mercury(I)		
Carbonate	$Hg_2CO_3 \rightleftharpoons Hg_2^{2+} + CO_3^{2-}$	8.9×10^{-17}
Chloride	$Hg_2Cl_2 \rightleftharpoons Hg_2^{2+} + 2Cl^-$	1.3×10^{-18}
Hydroxide	$Hg_2(OH)_2 \rightleftharpoons Hg_2^{2+} + 2OH^-$	1×10^{-23}
Sulfate	$Hg_2SO_4 \rightleftharpoons Hg_2^{2+} + SO_4^{2-}$	7.1×10^{-7}
Thiocyanate	$Hg_2(SCN)_2 \rightleftharpoons Hg_2^{2+} + 2SCN^-$	2.0×10^{-20}
Mercury(II)		
Hydroxide	$Hg(OH)_2 \rightleftharpoons Hg^{2+} + 2OH^-$	3.0×10^{-26}
Sulfide	$HgS \rightleftharpoons Hg^{2+} + S^{2-}$	1.0×10^{-52}
Nickel		
Carbonate	$NiCO_3 \rightleftharpoons Ni^{2+} + CO_3^{2-}$	6.6×10^{-9}
Hydroxide	$Ni(OH)_2 \rightleftharpoons Ni^{2+} + 2OH^-$	1×10^{-15}
Sulfide	$NiS \rightleftharpoons Ni^{2+} + S^{2-}$	1.0×10^{-22}
Silver		
Acetate	$AgC_2H_3O_2 \rightleftharpoons Ag^+ + C_2H_3O_2^-$	2.3×10^{-3}
Bromide	$AgBr \rightleftharpoons Ag^+ + Br^-$	5.0×10^{-13}
Carbonate	$Ag_2CO_3 \rightleftharpoons 2Ag^+ + CO_3^{2-}$	6.3×10^{-12}
Chloride	$AgCl \rightleftharpoons Ag^+ + Cl^-$	1.8×10^{-10}
Chromate	$Ag_2CrO_4 \rightleftharpoons 2Ag^+ + CrO_4^{2-}$	1.9×10^{-12}
Hydroxide	$AgOH \rightleftharpoons Ag^+ + OH^-$	1.9×10^{-8}
Iodide	$AgI \rightleftharpoons Ag^+ + I^-$	8.3×10^{-17}
Nitrite	$AgNO_2 \rightleftharpoons Ag^+ + NO_2^-$	1.6×10^{-4}
Oxalate	$Ag_2C_2O_4 \rightleftharpoons 2Ag^+ + C_2O_4^{2-}$	1.1×10^{-11}
Phosphate	$Ag_3PO_4 \rightleftharpoons 3Ag^+ + PO_4^{3-}$	2.0×10^{-21}
Sulfate	$Ag_2SO_4 \rightleftharpoons 2Ag^+ + SO_4^{2-}$	1.6×10^{-5}
Sulfide	$Ag_2S \rightleftharpoons 2Ag^+ + S^{2-}$	6.3×10^{-50}
Thiocyanate	$AgSCN \rightleftharpoons Ag^+ + SCN^-$	1.0×10^{-12}
Tin(II)		
Hydroxide	$Sn(OH)_2 \rightleftharpoons Sn^{2+} + 2OH^-$	1×10^{-25}
Sulfide	$SnS \rightleftharpoons Sn^{2+} + S^{2-}$	1.0×10^{-25}
Zinc		
Carbonate	$ZnCO_3 \rightleftharpoons Zn^{2+} + CO_3^{2-}$	6×10^{-11}
Hydroxide	$Zn(OH)_2 \rightleftharpoons Zn^{2+} + 2OH^-$	5×10^{-17}
Phosphate	$Zn_3(PO_4)_2 \rightleftharpoons 3Zn^{2+} + 2PO_4^{3-}$	1.0×10^{-32}
Sulfide	$ZnS \rightleftharpoons Zn^{2+} + S^{2-}$	1.6×10^{-23}

hydroxide, $Cd(OH)_2$. The equation for the reaction by which this substance is formed is

$$Cd^{2+} + 2\,OH^- \longrightarrow Cd(OH)_2(s)$$

Note that we always write an equation of this sort in the *net ionic* form, including *only* the species that change during the reaction. Since the Na^+ ion from the NaOH solution and the NO_3^- ion from the $Cd(NO_3)_2$ solution are still in the solution, essentially unchanged, when the system reaches equilibrium, they are *not* included in the equation.

The precipitation reaction proceeds to a state in which $Cd(OH)_2(s)$ is in equilibrium with its ions. If we wish, we can describe the equilibrium system by the equation

$$Cd(OH)_2(s) \rightleftharpoons Cd^{2+} + 2\,OH^-$$
$$K_{sp} = [Cd^{2+}][OH^-]^2 = 2 \times 10^{-14}$$

The equilibrium constant for this reaction is called the *solubility product* of $Cd(OH)_2$. (In Table 1-2 are listed the solubility products for many of the solids that will be studied in this course.) The meaning of the equilibrium condition for the preceding reaction is as follows:

In any aqueous solution in equilibrium with solid cadmium hydroxide, it must be true that $[Cd^{2+}] \times [OH^-]^2 = 2 \times 10^{-14}$.

Since K for this reaction is small, it should be apparent that, in a system in which $Cd(OH)_2$ is in equilibrium with its ions, either $[Cd^{2+}]$ or $[OH^-]$ must be small. They both *may* be small, which would be the situation if $Cd(OH)_2$ were dissolved in pure water. On the other hand, the concentration of one of the ions might be relatively large if that of the other ion were exceedingly small.

We must now consider two other matters related to the $Cd(OH)_2$ equilibrium system. The first concerns the conditions under which solid $Cd(OH)_2$ forms, and the second, the conditions under which this precipitate can be dissolved. It is certainly *not* true that mixing any two solutions containing Cd^{2+} and OH^- ions produces a precipitate. The condition that must be met in the combined solution just after mixing is

$$(\text{conc } Cd^{2+}) \times (\text{conc } OH^-)^2 > 2 \times 10^{-14}$$

If the concentrations of Cd^{2+} and OH^- are high enough so that this condition is met, equilibrium can, and will, be established by the precipitation of $Cd(OH)_2$; the concentrations of both ions decreases as the precipitation proceeds and decreases to such values that the previously mentioned product becomes equal to 2×10^{-14}. On the other hand, if the concentrations are so low that this product is $< 2 \times 10^{-14}$, equilibrium with $Cd(OH)_2$ cannot be established and no precipitate forms.

If we have a precipitate of $Cd(OH)_2$ that we wish to dissolve, we can apply the solubility product principle to choose a suitable reagent for the solution process. In this case, we start with an equilibrium system in which

$$[Cd^{2+}][OH^-]^2 = 2 \times 10^{-14}$$

Our objective is to add a reagent that will react with one of these ions, temporarily reducing its concentration *below* the value required for equilibrium. To restore equilibrium, it is necessary for some solid $Cd(OH)_2$ to dissolve, thereby bringing the concentration product back to 2×10^{-14}. One appropriate reagent would be an acid, such as 6M HCl, which contains H^+ ion that will react with the OH^- ion, produced by dissolving $Cd(OH)_2$, to form water and in the process decrease $[OH^-]$. Another useful reagent would be 6M NH_3, which would react with the Cd^{2+} to form $Cd(NH_3)_4{}^{2+}$ complex ion. Many precipitates, such as $Cd(OH)_2$, can be dissolved by addition of acids or complexing species to the solution with which they are in equilibrium. A more detailed treatment of these solution procedures will be given in the following two sections.

ACID-BASE REACTIONS

One of the most common reactions that occurs on mixing solutions is that between acids and bases. When a solution of a strong acid such as HCl, is mixed with that of a strong base, such as NaOH, a neutralization reaction occurs:

$$H^+ + OH^- \longrightarrow H_2O \qquad\qquad K = 1 \times 10^{14}$$

Since K is so large, the reaction goes essentially to completion, until one of the reacting species is used up. In the final solution, water is in equilibrium with its ions:

$$H_2O \rightleftharpoons H^+ + OH^- \qquad\qquad K_W = 1 \times 10^{-14}$$

The foregoing condition must be satisfied in all aqueous solutions, because water is always present. Since $[H^+]$ is by no means necessarily equal to $[OH^-]$, we can distinguish three types of solutions:

Neutral	$[H^+] = [OH^-] = 1 \times 10^{-7}$ M	
Acidic	$[H^+] > [OH^-]$	$[H^+] > 1 \times 10^{-7}$ M; $[OH^-] < 1 \times 10^{-7}$ M
Basic	$[H^+] < [OH^-]$	$[H^+] < 1 \times 10^{-7}$ M; $[OH^-] > 1 \times 10^{-7}$ M

Pure water and certain salt solutions are neutral. Acidic solutions are prepared by dissolving certain substances, such as HCl or $HC_2H_3O_2$, which contain an ionizable hydrogen atom, in water. Basic solutions can be made by dissolving solutes containing an ionizable OH^- ion in water. As we shall discuss later in this section, some salt solutions also exhibit definite acidic or basic properties. When solutions having acidic and basic characteristics (regardless of source) are mixed, the neutralization reaction proceeds until the condition of the equilibrium between water and its ions is satisfied.

Let us now return to the $Cd(OH)_2$ solubility problem and treat it somewhat more quantitatively than we could previously. Since $Cd(OH)_2$, or any insoluble hydroxide, tends to furnish some hydroxide ion to a solution, making it basic, we might expect that the solubility of such a substance would depend very markedly on the acidity or basicity of the solution we use as solvent. For example, suppose that we add some solid $Cd(OH)_2$ to a liter of 0.1M NaOH. How many moles would dissolve? The equation for the reaction would be

$$Cd(OH)_2(s) \rightleftharpoons Cd^{2+} + 2\,OH^- \qquad K_{sp} = [Cd^{2+}][OH^-]^2 = 2 \times 10^{-14}$$

Since the equilibrium condition must be satisfied, and since initially $[Cd^{2+}]$ equals zero, some cadmium hydroxide does dissolve. Initially, however, in the basic solution, $[OH^-]$ equals 0.1M, and as soon as $[Cd^{2+}]$ increases to 2×10^{-12} M, it becomes true that

$$[Cd^{2+}][OH^-]^2 = (2 \times 10^{-12})(0.1)^2 = 2 \times 10^{-14}$$

and the equilibrium condition is met, with only 2×10^{-12} moles of $Cd(OH)_2$ having gone into solution. (The amount of OH^- contributed by $Cd(OH)_2$ is negligible compared to that from the NaOH.) Therefore, not much Cd(OH) dissolves; from this calculation we can conclude that (1) $Cd(OH)_2$ is for all practical purposes insoluble in 0.1M NaOH and (2) if we precipitate Cd^{2+} in a strongly basic solution in which $[OH^-]$ is 0.1M, essentially all the Cd^{2+} will be precipitated as $Cd(OH)_2$. The same situation would exist in solutions much less basic than 0.1M NaOH; even in a very slightly basic solution, where $[OH^-]$ is 1×10^{-4} M, $[Cd^{2+}]$ at equilibrium cannot be greater than 2×10^{-6} M, or about 0.1 mg/liter. For this reason, $Cd(OH)_2$ and many other hydroxides are not appreciably soluble in basic solutions.

Now let us examine the solubility of $Cd(OH)_2$ in a neutral solution, that is, a solution in which $[H^+]$ equals $[OH^-]$ equals 1×10^{-7} M. Under these conditions, the equilibrium concentration of cadmium ion would be given by

$$[Cd^{2+}] = \frac{2 \times 10^{-14}}{(1 \times 10^{-7})^2} = 2M$$

This calculation tells us that 2 moles, roughly 300 g, of $Cd(OH)_2$ can dissolve in 1 liter of solution in which the *equilibrium* concentration of H^+ is 1×10^{-7} M.

In practical terms, we see that a very large amount of $Cd(OH)_2$ can be dissolved by adding sufficient acid to give a neutral solution at equilibrium.

For example, if we had 0.1 mole of $Cd(OH)_2$ in 1 liter of water, and we wished to dissolve that material completely, we might add acid, drop by drop, to the water. Before we added any acid, the equilibrium concentration of OH^- would be roughly 3×10^{-5} M, because of dissolved $Cd(OH)_2$. As we added H^+ it would react with the OH^- present in solution, decreasing its concentration and forcing more $Cd(OH)_2$ to dissolve to maintain equilibrium. Adding 0.2 mole of H^+ would furnish enough acid to the solution to react with any OH^- ion available from dissolving hydroxide, and would leave the solution essentially neutral, i.e., $[OH^-] = 1 \times 10^{-7}$ M. At that point, $[Cd^{2+}]$ would be 0.1M, and the concentration product, $(\text{conc } Cd^{2+})(\text{conc } OH^-)^2$, would be less than 2×10^{-14}; all the $Cd(OH)_2$ would dissolve and there would be no equilibrium between the solid and its ions.

In most cases in qualitative analysis, when we set out to dissolve an hydroxide precipitate, we add more than enough acid to react with all the OH^- available from the solid, ensuring that the final solution will be acidic. Under such conditions, any hydroxide precipitate will be completely soluble, since the value of $[OH^-]$ in an acid solution is so low that even dissolving large amounts of the hydroxide cannot furnish enough cation to increase the concentration product to the value of K_{sp}.

Weak Acids and Bases. Solutions of HCl, HBr, HNO_3, and $NaOH$ are very highly ionized; therefore, in 1M HCl, for example, we can say that $[H^+]$ equals 1M, and in 0.5M NaOH, $[OH^-]$ equals 0.5M. A substance like HCl is called a strong acid, and one like NaOH is called a strong base, because in water solutions their ionization is nearly 100 per cent; in an HCl solution there are essentially no HCl molecules, and an NaOH solution contains effectively no NaOH molecules.

Many acids differ from HCl in that they only slightly dissociate in water solution. Rather than being about 100 per cent ionized at moderate concentrations, they may be only 10, 1, or 0.1 per cent ionized. These acids are called weak acids; they are far more numerous than strong acids.

We will encounter many weak acids in this course, including carbonic acid, acetic acid, sulfurous acid, hydrogen sulfide, phosphoric acid, and oxalic acid. Both the weak acids and their salts have properties that depend in important ways on the law of chemical equilibrium.

Acetic acid, $HC_2H_3O_2$, which we will abbreviate to HA, is a typical weak acid. In aqueous solution it dissociates according to the equation:

$$HA \rightleftharpoons H^+ + A^-$$

$$K_a = \frac{[H^+][A^-]}{[HA]} = 2 \times 10^{-5}$$

The equilibrium constant for the reaction, K_a, is called the dissociation or ionization constant of acetic acid. In Table 1-3 are listed the ionization constants of the weak acids of the anions included in the qualitative analysis scheme. Since K_a is usually small, the ionization reaction does not go far to the right. By calculation, we can show that 0.1M acetic acid is just about 1 per cent ionized. (To check our calculation, note that if the acid is 1 per cent ionized, $[H^+]$ and $[A^-]$ must both be 1×10^{-3} M, and, since only a little HA ionizes, [HA] is still just about 0.1M. Substituting into the expression for K_a, you can see that the quotient is about 10^{-5}.)

Weak acids, like strong ones, react with bases to form salts. On addition of a solution of NaOH to acetic acid, we would have the reaction:

$$HA + OH^- \longrightarrow A^- + H_2O \qquad\qquad K = 2 \times 10^9$$

In the net ionic equation we write HA rather than H^+, since in the acid solution the species that predominates is the HA molecule. The value of K for this reaction can be easily derived, since the reaction is the sum of two other reactions we have already discussed:

I $HA \rightleftharpoons H^+ + A^-$ $\qquad\qquad K_I = K_a = 2 \times 10^{-5}$
II $H^+ + OH^- \rightleftharpoons H_2O$ $\qquad\qquad K_{II} = 1/K_w = 1 \times 10^{14}$
III $HA + OH^- \rightleftharpoons A^- + H_2O$ $\qquad\qquad K_{III} = K_I \times K_{II} = 2 \times 10^9$

TABLE 1-3.   **Ionization Constants of Weak Acids and Bases at 25° C**

	Acid	K_a	*Base*	K_b
Acetic acid	$HC_2H_3O_2$	1.8×10^{-5}	$C_2H_3O_2^-$	5.6×10^{-10}
Ammonium ion	NH_4^+	5.6×10^{-10}	NH_3	1.8×10^{-5}
Carbonic acid	H_2CO_3	4.2×10^{-7}	HCO_3^-	2.4×10^{-8}
	HCO_3^-	4.8×10^{-11}	CO_3^{2-}	2.1×10^{-4}
Chromic acid (K_2)	$HCrO_4^-$	3.2×10^{-7}	CrO_4^{2-}	3.2×10^{-8}
Hydrocyanic acid	HCN	4.0×10^{-10}	CN^-	2.5×10^{-5}
Hydrogen sulfide	H_2S	1×10^{-7}	HS^-	1×10^{-7}
	HS^-	1×10^{-15}	S^{2-}	1×10^1
Nitrous acid	HNO_2	4.5×10^{-4}	NO_2^-	2.2×10^{-11}
Oxalic acid	$(COOH)_2$	5.6×10^{-2}	$COOHCOO^-$	1.6×10^{-13}
		5.2×10^{-5}	$(COO)_2^{2-}$	1.9×10^{-10}
Phosphoric acid	H_3PO_4	7.5×10^{-3}	$H_2PO_4^-$	1.3×10^{-12}
	$H_2PO_4^-$	6.2×10^{-8}	HPO_4^{2-}	1.6×10^{-7}
	HPO_4^{2-}	1.7×10^{-12}	PO_4^{3-}	5.9×10^{-3}
Sulfuric acid (K_2)	HSO_4^-	1.0×10^{-2}	SO_4^{2-}	1.0×10^{-12}
Sulfurous acid	H_2SO_3	1.7×10^{-2}	HSO_3^-	5.9×10^{-13}
	HSO_3^-	5.6×10^{-8}	SO_3^{2-}	1.8×10^{-7}

For the acid HA, K_a is the equilibrium constant for the reaction:

$$HA \rightleftharpoons H^+ + A^- \qquad\qquad K_a = \frac{[H^+][A^-]}{HA}$$

For the base A^-, K_b is the equilibrium constant for the reaction:

$$A^- + H_2O \rightleftharpoons HA + OH^- \qquad\qquad K_b = \frac{[HA][OH^-]}{[A^-]}$$

For a given acid-base system, it is true that $\qquad\qquad K_b = \dfrac{K_w}{K_a}$

Since K_{III} is very large, acetic acid, and most weak acids, react very nearly completely with strong bases when solutions of the two are mixed. The product of the reaction is a salt, NaA, if NaOH is the base used. The NaA ionizes completely in solution, yielding Na^+ and A^-, as is really implied by the net ionic equation we wrote for the neutralization reaction.

Given a solution of sodium acetate, NaA, it is easy to see that it must behave as a weak base. In such a solution the Na^+ ion is neutral, since it does not tend to react with water. However, the A^- tends to react to some extent with water, since the reverse of reaction III must occur to satisfy the equilibrium law:

$$A^- + H_2O \rightleftharpoons HA + OH^- \qquad K_b = 5 \times 10^{-10} = \frac{[HA][OH^-]}{[A^-]}$$

Physically, this reaction occurs because HA is a weak acid, and A^- ion tends to a slight extent to extract H^+ ions from water to make HA molecules. The solution of NaA becomes alkaline, since it contains a greater concentration of OH^- ion than H^+ ion. The extent of the reaction is very slight; in a 0.1M solution of sodium acetate, one can show that $[OH^-]$ must be about 7×10^{-6} M, just on the basic side.

Salts of weak acids in general form slightly basic solutions; these salts are weak bases, since they produce a small amount of OH^- ion on reaction with water. They are indeed the most common weak bases.

One weak base that is often used is not the salt of a weak acid. This base is a water solution of NH_3, ammonia. Its solution is basic because of the reaction:

$$NH_3 + H_2O \rightleftharpoons NH_4^+ + OH^- \qquad K_b = 2 \times 10^{-5} = \frac{[NH_4^+][OH^-]}{[NH_3]}$$

In a 0.1M NH_3 solution, the concentration of OH^- is just about 1×10^{-3} M; therefore roughly 1 per cent of the NH_3 reacts to form NH_4^+ and OH^-. Although NH_3 is a gas and is quite soluble in water, it can be driven from a basic solution by simply boiling it out. Ammonium salts are weak acids and react with strong bases by the reverse of the previously mentioned reaction.

A mixture of an ammonium salt and ammonia acts as a *buffer* solution; a buffer resists change of its acidity or basicity much more than does a normal solution. For example, a mixture of 1M NH_3 and 1M NH_4Cl, in which $[NH_3]$ and $[NH_4^+]$ both equal approximately 1M, obeys the equilibrium law; from the condition on the reaction of NH_3 with water, just mentioned, you can see that in that solution $[OH^-]$ must be 2×10^{-5} M. Addition of even 0.2 mole of H^+ or 0.2 mole of OH^- does not change $[OH^-]$ very much, since the added H^+ is removed by reaction with NH_3, and the OH^- is removed by reaction with NH_4^+. (Both NH_3 and NH_4^+ are present in sufficient excess so that the actual amounts of these two species do not change much as a result of their reacting with the added acid or base.)

Dissolving Salts of Weak Acids. Nearly all the salts of the weak acids are insoluble in water, the only common exceptions being the acetates and the salts of the alkali metals and the ammonium ion. These salts exhibit some interesting acid-base properties of importance in qualitative analysis.

Consider zinc carbonate, $ZnCO_3$, which is a typical salt of a weak acid, in this case carbonic acid. The carbonate dissolves slightly in water, the equation for the reaction being:

$$ZnCO_3(s) \longrightarrow Zn^{2+} + CO_3^{2-} \qquad\qquad K_{sp} = 6 \times 10^{-11}$$

Since K_{sp} is small, only a little $ZnCO_3$ goes into solution in water, the saturated solution being about 8×10^{-6} M.

Frequently in qualitative analysis we need to dissolve an insoluble salt, such as $ZnCO_3$, so that we can then analyze it for the cation or anion it contains. Our usual procedure is to add a reagent that disturbs the equilibrium system by decreasing the concentration of one of the ions, making the concentration product, in this case, (conc Zn^{2+}) (conc CO_3^{2-}), less than K_{sp}, and thereby forcing the solid to dissolve to satisfy the equilibrium condition. With all the salts of weak acids, as with the hydroxides, the addition of H^+ ion creates just such a disturbance.

The equation for the reaction of $ZnCO_3$ with acid is best written as:

$$\text{IV} \quad ZnCO_3(s) + 2\,H^+ \rightleftharpoons Zn^{2+} + CO_2 + H_2O$$

This reaction is the sum of several other reactions:

I	$ZnCO_3(s) \rightleftharpoons Zn^{2+} + CO_3^{2-}$	$K_I = K_{sp} = 6 \times 10^{-11}$
II	$H^+ + CO_3^{2-} \rightleftharpoons HCO_3^-$	$K_{II} = 1/K_2 = 2 \times 10^{10}$
III	$H^+ + HCO_3^- \rightleftharpoons CO_2 + H_2O$	$K_{III} = 1/K_1 = 2.5 \times 10^6$

Reactions II and III are the reverse of the reactions by which carbonic acid, which is a solution of carbon dioxide in water, ionizes:

$$H_2CO_3 = H_2O + CO_2 \rightleftharpoons H^+ + HCO_3^- \qquad K_1 = 5 \times 10^{-11}$$
$$HCO_3^- \rightleftharpoons H^+ + CO_3^{2-} \qquad K_2 = 4 \times 10^{-7}$$

Since Reaction IV is the sum of Reactions I, II, and III, it follows that

$$K_{IV} = K_I K_{II} K_{III} = \frac{K_{sp}}{K_1 K_2} = \frac{[Zn^{2+}][CO_2]}{[H^+]^2} = 3 \times 10^6$$

The large value of K_{IV} implies that $ZnCO_3$ reacts strongly with acid and dissolves. In a solution in which the solid carbonate is in equilibrium with its ions and $[H^+]$ equals 1×10^{-4}, only slightly acidic, since by reaction IV $[Zn^{2+}]$ equals $[CO_2]$, we can see that

$$[Zn^{2+}]^2 = 3 \times 10^6 \times (1 \times 10^{-4})^2 = 3 \times 10^{-2}$$
$$Zn^{2+} = 0.17M$$

At higher values of H^+, much more carbonate would go into solution.

As with hydroxides, when attempting to dissolve salts such as $ZnCO_3$ in qualitative analysis procedures, we usually add enough acid to be sure that even when the solid is dissolved the solution will still be quite acidic. Under such conditions nearly all salts of the weak acids are completely soluble. Sometimes, as with the carbonates, the weak acid is a solution of a gas in water; in the acid solution the concentration of the gas may become so great that gas bubbles may leave the solution. Acidification of carbonates, sulfites, and sulfides often produces such an effervescence.

The properties of $ZnCO_3$ in the presence of acids explain our observations in the qualitative analysis problem discussed earlier. There you will recall that addition of Na_2CO_3 to the acidic solution of Zn^{2+} did not precipitate $ZnCO_3$ immediately, but rather produced gas bubbles. In that solution, addition of carbonate ion caused the following reaction to occur:

$$2 H^+ + CO_3^{2-} \rightleftharpoons CO_2 + H_2O \qquad K = \frac{1}{K_1 K_2} = 5 \times 10^{16}$$

This reaction went essentially to completion, increasing $[CO_2]$ until carbon dioxide bubbles formed; as long as the solution remained acidic, $[CO_3^{2-}]$ was kept so small that $ZnCO_3$ could not precipitate. Finally, when all the H^+ ion was used up, further addition of Na_2CO_3 produced an appreciable concentration of CO_3^{2-} ion and brought down the solid zinc carbonate.

One important group of salts of weak acids do not dissolve in solutions of nonoxidizing acids. There are the sulfides of some of the transition metals, including CuS, PbS, HgS, and CdS. These substances are extremely insoluble in water. If we consider CdS, and write the equation for its reaction with acid:

$$\text{IV} \quad CdS(s) + 2 H^+ \rightleftharpoons Cd^{2+} + H_2S$$

As with the $ZnCO_3$ system, this equation can be written as a sum:

$$\text{I} \quad CdS(s) \rightleftharpoons Cd^{2+} + S^{2-} \qquad K_I = K_{sp} = 1 \times 10^{-26}$$
$$\text{II} \quad H^+ + S^{2-} \rightleftharpoons HS^- \qquad K_{II} = 1/K_2 = 1 \times 10^{15}$$
$$\text{III} \quad H^+ + HS^- \rightleftharpoons H_2S \qquad K_{III} = 1/K_1 = 1 \times 10^7$$

where K_1 and K_2 are the first and second ionization constants, respectively, of H_2S. Since reaction IV is the sum of the three reactions:

$$K_{IV} = \frac{K_{sp}}{K_1 K_2} = 1 \times 10^{-4} = \frac{[Cd^{2+}][H_2S]}{[H^+]^2}$$

By the preceding relationship we can see that in a solution of 0.1M HCl, the concentration of Cd^{2+} in equilibrium with CdS is low, of the order of magnitude of 1×10^{-5} M if the solution is saturated (0.1M) with H_2S. This means that CdS does not readily go into solution in an acid and that addition of H_2S to a moderately acidic solution of Cd^{2+} precipitates CdS. In one part of the qualitative analysis scheme, the difference in solubilities of various sulfides in acid solution is used to separate one group of cations from another.

COMPLEX ION FORMATION— AMPHOTERISM

Although we write the formulas of the cations as though they exist as bare species in water solution, such an ion often is a more complex species, in which the cation is bonded to several water molecules. The Cr^{3+} ion in water solution may have the formula $Cr(H_2O)_6^{3+}$. These hydrated ions are examples of complex ions; a *complex ion* consists of a central metal ion to which are bonded, often somewhat loosely, two, four, or six neutral or ionic species, called *ligands*. Some of the common ligands are H_2O, NH_3, OH^-, CN^-, S^{2-}, and Cl^-. When the ligand is H_2O it is ordinarily omitted from the formula for the ion; in this book, the formulas of the cations usually are written as though the ion were bare unless the explanation for a reaction involves the formation or decomposition of a specific complex ion.

A complex ion ordinarily exists in a state of equilibrium with the cation and ligands into which it dissociates. The stability of a complex ion is reflected in the equilibrium constant for the dissociation reaction. For the cadmium ammonia complex ion the reaction is written as:

$$Cd(NH_3)_4^{2+} \rightleftharpoons Cd^{2+} + 4\,NH_3 \qquad\qquad K = 1 \times 10^{-7}$$

In Table 1-4 are listed the dissociation constants of some of the common complex ions. For many of the transition metal cations, the stabilities of the complex ions formed by reaction with different ligands usually increase in the following order: chloride < ammonia < cyanide. The relative stabilities of complex ions with different ligands may differ somewhat with the cation under study; the fact that a cation forms a complex ion with one ligand does not necessarily imply that it forms a complex ion with every other common ligand. Cadmium ion, for example, forms an ammonia complex ion, but it does not readily form a stable hydroxide complex ion.

In the example of a simple analysis that we used at the beginning of the chapter, we formed two complex ions when we added NaOH to the cation mixture. The reaction involving the zinc ion was

$$Zn^{2+} + 4\,OH^- \longrightarrow Zn(OH)_4^{2-} \qquad\qquad K = 1/K_{diss} = 3 \times 10^{15}$$

TABLE 1-4.   **Dissociation Constants of Some Complex Ions at 25° C**

Species	Reaction	K_{diss}
Aluminum		
Hydroxide	$Al(OH)_4^- \rightleftharpoons Al^{3+} + 4\ OH^-$	5×10^{-34}
Cadmium		
Ammonia	$Cd(NH_3)_4^{2+} \rightleftharpoons Cd^{2+} + 4\ NH_3$	1.2×10^{-7}
Iodide	$CdI_6^{4-} \rightleftharpoons Cd^{2+} + 6\ I^-$	1×10^{-6}
Cobalt(II)		
Ammonia	$Co(NH_3)_6^{2+} \rightleftharpoons Co^{2+} + 6\ NH_3$	1.6×10^{-5}
Cobalt(III)		
Ammonia	$Co(NH_3)_6^{3+} \rightleftharpoons Co^{3+} + 6\ NH_3$	6×10^{-36}
Copper(II)		
Ammonia	$Cu(NH_3)_4^{2+} \rightleftharpoons Cu^{2+} + 4\ NH_3$	2.5×10^{-13}
Oxalate	$Cu(C_2O_4)_2^{2-} \rightleftharpoons Cu^{2+} + 2\ C_2O_4^{2-}$	5×10^{-11}
Iron(III)		
Oxalate	$Fe(C_2O_4)_3^{3-} \rightleftharpoons Fe^{3+} + 3\ C_2O_4^{2-}$	6×10^{-21}
Mercury(II)		
Chloride	$HgCl_4^{2-} \rightleftharpoons Hg^{2+} + 4\ Cl^-$	6×10^{-16}
Iodide	$HgI_4^{2-} \rightleftharpoons Hg^{2+} + 4\ I^-$	6×10^{-31}
Nickel		
Ammonia	$Ni(NH_3)_6^{2+} \rightleftharpoons Ni^{2+} + 6\ NH_3$	3×10^{-9}
Silver		
Ammonia	$Ag(NH_3)_2^+ \rightleftharpoons Ag^+ + 2\ NH_3$	4×10^{-8}
Zinc		
Ammonia	$Zn(NH_3)_4^{2+} \rightleftharpoons Zn^{2+} + 4\ NH_3$	8×10^{-10}
Hydroxide	$Zn(OH)_4^{2-} \rightleftharpoons Zn^{2+} + 4\ OH^-$	3×10^{-16}

The zinc hydroxide complex ion, for which the official name is the tetra-hydroxozinc(II) ion, is very stable; in a solution that is 1M in OH^- ion, the ratio $[Zn(OH)_4^{2-}]/[Zn^{2+}]$ is of the order of 3×10^{15}, which means that just about all the zinc in such a solution is in the form of the complex ion. Zinc hydroxide, $Zn(OH)_2$, is actually very insoluble (K_{sp} equals 5×10^{-17}), but in the presence of appreciable amounts of hydroxide ion it readily dissolves:

$$Zn(OH)_2(s) + 2\ OH^- \rightleftharpoons Zn(OH)_4^{2-} \qquad K = 0.15$$

(The K for this reaction can be derived by the usual procedure from the value of K_{sp} for the hydroxide and the dissociation constant of the complex ion.) The value of K tells us that in 1 liter of 1M NaOH we can dissolve about 0.1 mole of $Zn(OH)_2$. Can you see why? In 6M NaOH, any reasonable amount of $Zn(OH)_2$ is completely soluble.

The lead ion, like the zinc ion, forms an hydroxide complex with the probable formula $Pb(OH)_4^{2-}$. The hydroxides of lead and zinc are soluble in both basic and acidic solutions; hydroxides having this property are some-times called *amphoteric*. This term is of more historic than scientific impor-tance; hydroxide complex ions are completely analogous to other complex ions in their equilibrium behavior, and there is no valid reason to distinguish them. We will use the term amphoteric occasionally, mainly because it offers a way to indicate with a word that the cation in question forms an hydroxide complex ion.

Formation of complex ions is often used to dissolve precipitates that do not easily go into solution in acids. For example, AgCl and $PbSO_4$ are two insoluble salts. Since they are both salts of strong acids, they do not tend to dissolve in acids; the Cl^- and SO_4^{2-} ions do not interact appreciably with H^+ ion, a necessary condition if acid is to be an effective solvent. However, Ag^+ does form a moderately stable ammonia complex ion:

$$Ag(NH_3)_2^+ \rightleftharpoons Ag^+ + 2\,NH_3 \qquad\qquad K_{diss} = 4 \times 10^{-8}$$

Silver chloride, ($K_{sp} = 2 \times 10^{-10}$) may dissolve in ammonia solutions by the reaction:

$$AgCl(s) + 2\,NH_3 \rightleftharpoons Ag(NH_3)_2^+ + Cl^- \qquad K = K_{sp}/K_{diss} = 5 \times 10^{-3}$$

Given the value of K for this reaction, it is easy to show that in 1M NH_3 we can expect to dissolve about 0.07 mole of AgCl/liter. At higher NH_3 concentrations, AgCl would be even more soluble. Neither AgBr nor AgI, being less soluble than AgCl, readily goes into solution in NH_3.

Like silver chloride, lead sulfate is soluble in some complexing reagents; for example, it dissolves in strong, hot NaOH solution.

Reactions involving complex ion formation are used extensively in qualitative analysis. Given a precipitate that may include a rather large group of cations, one can selectively remove some of those cations from the precipitate by adding a complexing reagent that dissolves one or more of the solids present, putting some of the cations into solution and leaving others in the solid precipitate. After separation of the solid from the liquid, one can usually destroy the complex ions in the solution by adding the proper reagent; this may or may not reprecipitate the cations that were dissolved.

A simple example of how this approach works very nicely is in the scheme of analysis of the Group I cations. Assume that we have a precipitate that may contain AgCl and Hg_2Cl_2. By adding 6M NH_3 to the solid we form the $Ag(NH_3)_2^+$ complex ion and thereby dissolve the AgCl; since Hg_2Cl_2 reacts with, but is not dissolved by, NH_3, any Hg_2^{2+} remains in a solid phase. After the solution containing $Ag(NH_3)_2^+$ has been separated from the solid, the complex ion is caused to decompose by treatment with acid. The H^+ ion reacts with NH_3 in equilibrium with the complex, forming NH_4^+; this reaction causes $[NH_3]$ to decrease. The complex ion must dissociate to meet the equilibrium condition; as the complex breaks down, $[Ag^+]$ increases, ultimately reaching a value that causes AgCl to reprecipitate. If it were not possible to complex the silver ion in this way, the distinction between Ag^+ and Hg_2^{2+} would be much more difficult to establish.

OXIDATION-REDUCTION REACTIONS

Most of the reactions encountered in qualitative analysis can be classified as simple or combined precipitation, acid-base, and complex ion reactions. The analysis scheme is based on selective precipitation and solution reactions, and for most purposes the kinds of reactions we have already discussed will suffice.

However, there are a few instances in which it is necessary to resort to yet another kind of reaction, involving oxidation or reduction, to accomplish a desired change. The first example of such a situation is when we are faced with dissolving the sulfides of copper, cadmium, bismuth, and lead in an early stage of the analysis of the Group II cations. As we noted previously, these substances are so insoluble that they do not go into solution in acids as any properly behaving salt of a weak acid should. Neither will these solids dissolve in any useful complexing agent. To render these salts tractable, we may either oxidize the sulfide they contain to the elementary state or reduce the cations they contain to the metallic state. Thereby we can decompose the salt and ultimately get the cations back into solution.

In the case in point, we choose to oxidize the sulfide to free sulfur, using 6M HNO_3 as the oxidizing reagent. Hot nitric acid, in the presence of oxidizable species, undergoes the following half reaction:

$$3\,e^- + 4\,H^+ + NO_3^- \longrightarrow NO + 2\,H_2O$$

Some NO_2 may also form, but in any event the reaction is one requiring electrons, which are furnished by the oxidized species. With the sulfides, the oxidation reaction for Bi_2S_3 is:

$$Bi_2S_3(s) \longrightarrow 2\,Bi^{3+} + 3\,S(s) + 6\,e^-$$

If we take account of the fact that no electrons can be produced or consumed, the overall oxidation-reduction reaction is:

$$8\,H^+ + 2\,NO_3^- + Bi_2S_3(s) \longrightarrow 2\,NO + 4\,H_2O + 2\,Bi^{3+} + 3\,S(s)$$

The reaction is carried out in hot solution, so that most of the NO is lost as a gas. The solution obtained is equivalent to that formed by dissolving bismuth nitrate in nitric acid; some solid sulfur is present and can be removed by centrifugation. Any excess HNO_3 can be removed by boiling; consequently, on treating CuS, Bi_2S_3, CdS, and PbS with nitric acid, one obtains the nitrate salts of those metals, an optimal result indeed.

A few sulfides are not dissolved by 6M HNO_3, including HgS, CoS, and NiS. These substances are taken into solution by *aqua regia,* a mixture of concentrated nitric and hydrochloric acids. This combines the good oxidizing power of HNO_3 with the complexing power of HCl. For HgS the overall reaction is:

$$3\,HgS(s) + 12\,Cl^- + 2\,NO_3^- + 8\,H^+ \longrightarrow$$
$$3\,HgCl_4^{2-} + 2\,NO + 4\,H_2O + 3\,S(s)$$

In addition to the treatment of sulfides, oxidizing reagents are useful in various other key situations in the analysis. Besides 6M HNO_3, which is probably the most common oxidizing agent in qualitative analysis, H_2O_2, BiO_3^-, $HgCl_2$, 18M H_2SO_4, and MnO_4^- are used at one or more points in the procedure; each of these species contains a reducible atom and is often selected in a particular case both to accomplish the required oxidation and to provide a specific analytical test. The half reactions for these agents are:

$$2\,e^- + H_2O_2 \longrightarrow 2\,OH^-$$
$$2\,e^- + 6\,H^+ + BiO_3^- \longrightarrow Bi^{3+} + 3\,H_2O$$
$$2\,e^- + 2\,HgCl_2 \longrightarrow Hg_2Cl_2(s) + 2\,Cl^-$$
$$2\,e^- + 2\,H^+ + H_2SO_4 \longrightarrow SO_2 + 2\,H_2O$$
$$5\,e^- + 8\,H^+ + MnO_4^- \longrightarrow Mn^{2+} + 4\,H_2O$$

Reducing agents are employed in a few steps in the qualitative analysis scheme, either to remove a cation from solution by reducing it to the metal or to lower its oxidation state for easier identification. Hydrogen gas generated in situ by reaction of Al with acid, Sn(II) in both acidic and basic solutions, and H_2O_2 in acidic solution are among the reducing species used. Each of these reagents contains an oxidizable atom; the half reactions that occur are:

$$H_2(g) \longrightarrow 2\,H^+ + 2\,e^-$$
$$Sn(II) \longrightarrow Sn(IV) + 2\,e^-$$
$$H_2O_2 \longrightarrow O_2(g) + 2\,H^+ + 2\,e^-$$

Although oxidation-reduction reactions in principle are subject to the law of chemical equilibrium, those that we carry out in the laboratory have large equilibrium constants, and for practical purposes the reactions proceed to completion. This kind of reaction differs from precipitation, acid-base, and complex ion formation reactions in that it is often relatively complicated and essentially irreversible. For these reasons and because they often involve species that contaminate the system, oxidizing and reducing agents are mainly employed either in final identification tests, or, as with the sulfides, in bringing otherwise inert substances into solution.

PROBLEMS *

1. Explain why $Cu(OH)_2$ is very soluble in 0.1M HCl but is essentially insoluble in 0.1M NaOH.

2. If K_a for acetic acid is 2×10^{-5}, calculate $[H^+]$ in 0.1M acetic acid. What is the percentage ionization of $HC_2H_3O_2$ in this solution?

3. Calculate $[OH^-]$ in a buffer in which $[NH_3]$ is 0.1M and $[NH_4^+]$ is 0.4M.

4. About how many moles of $ZnCO_3$ could be dissolved in 1 liter of solution in which the equilibrium concentration of H^+ is 1×10^{-3} M? Assume that $[Zn^{2+}]$ equals $[CO_2]$ in that solution.

5. Explain why $PbCO_3$ is soluble in 1M HNO_3 but PbS is not.

6. How many moles of $Zn(OH)_2$ would dissolve in 1 liter of 1M NaOH?

101. It is possible to dissolve $Mg(OH)_2$ very easily in acid solutions, even if they are dilute. Explain.

102. We note that in 0.1M NH_3, about 1 per cent of the ammonia reacts with water to form OH^- and NH_4^+. Assuming that K_b for NH_3 is 2.0×10^{-5}, calculate the exact percentage that reacts.

103. In a buffer containing 1M NH_3 and 1M NH_4Cl, $[OH^-]$ is 2×10^{-5} M. Find $[OH^-]$ in 1 liter of such a buffer after 0.1 mole of NaOH has been added.

104. About how many moles of $CaCO_3$ could be dissolved in 1 liter of 0.2M HCl?

105. Silver carbonate is soluble in 1M HNO_3 but silver chloride is not. Explain.

106. If we wish to dissolve 0.1 mole of AgCl in 1 liter of NH_3 solution, what is the minimal initial concentration of NH_3 that is necessary? What would be $[NH_3]$ in the final equilibrium solution?

---

* Throughout this text the problems at the end of each chapter are arranged in two sets, beginning with the numbers 1 and 101, respectively. Problem 1 is like 101, 2 is like 102, and so forth. Answers to problems numbered 101 and above are given in Appendix 2.

Certain problems on this page require quantitative answers. The mathematical techniques necessary for treating these problems were summarized in this chapter and are discussed in greater detail in texts in general chemistry. Problems in the other chapters of this text are all of a qualitative nature.

CHAPTER 2

LABORATORY PROCEDURES

In this course we carry out our analyses on what is commonly called the semimicro level. This means simply that the amounts of chemicals we use are neither large, of the order of grams, nor very small, of the order of micrograms. The sample solutions are typically about 0.1M, and one works with volumes on the order of 1 ml, which means that for an average solute ion there will be about 10 mg in the sample. The identification tests for the most part are effective if more than 1 or 2 mg of the species is present; therefore, you must carry out your procedures carefully enough to avoid losing the major portion of any component somewhere during the analysis.

EQUIPMENT USED IN SEMIMICRO ANALYSIS

There are several real advantages in performing analyses at the semimicro level. Most important, separations of solid precipitates from solutions can be accomplished rapidly and quantitatively by centrifuging rather than by filtering. Small laboratory centrifuges are commonly available and are the main large item of equipment to be used. These centrifuges work well with a semimicro test tube holding about 8 ml; the commercial designation of these tubes is 13 mm × 100 mm, and they will be used almost exclusively in all our work. Only very rarely will it be necessary, or helpful, to use a large test tube.

In your laboratory locker you should have about 10 semimicro test tubes and a test tube rack to hold them. You will also find the following items very useful:

3 glass stirring rods (4 mm) about 5 or 6 inches long, fire polished on both ends. These can be made from glass rod available in the laboratory.

4 medicine droppers	1 beaker (400 ml)
1 small metal spatula	2 beakers (150 ml)
1 graduated cylinder (10 ml)	2 beakers (30 ml)
1 plastic squeeze bottle for distilled water	1 bunsen burner

1 test tube holder
1 test tube brush
1 watch glass (6 cm)

1 wire gauze and ring stand
red and blue litmus paper
1 pair safety glasses

In addition to these items, you may be furnished with a set of small dropping bottles for the most common solutions. Chemical reagents will ordinarily be dispensed from 250 ml dropping bottles located on your laboratory bench.

COMMON LABORATORY OPERATIONS

As you perform analyses in the laboratory, you will find that certain procedures are used again and again. These procedures are really very simple and can easily be done properly by beginning students. As you may surmise, however, the most efficient manner of executing a given operation is not necessarily intuitively obvious. In the following discussion we will consider several of the laboratory procedures that are used in qualitative analysis and the manner in which they should be carried out.

Dispensing a Given Volume of Reagent Solution. A typical step in a procedure is to add 1 ml of a reagent to a solution. This is almost *never* done by adding the reagent to a graduated cylinder, measuring out 1 ml, and then pouring that into the sample. The direction given means to add about 1 ml, e.g., ±0.1 ml. This is easily done if you know approximately the volume 1 ml occupies in a semimicro test tube. The way to determine this is to measure out about 5 ml of water in a graduated cylinder and then to pour that into a test tube, 1 ml at a time. After you have done this a few times, you can accurately judge the increase in level of liquid corresponding to 1 ml, and indeed to 0.5 ml. In general use the medicine dropper in the reagent bottle to transfer solution directly to the test tube containing your sample. Use that dropper and *not* your own for dispensing reagents. *Never* contaminate reagent bottles by dipping eye droppers or spatulas into them. Distilled water from the squeeze bottle is added directly, with volume again estimated by the change in the level of liquid.

Since most medicine droppers deliver roughly the same volume per drop, it is sometimes helpful to count drops to establish volumes. You should check your own medicine droppers to see how many drops equal 1 ml; you can assume that those in the dropping bottles deliver about the same volume per drop. Typical droppers deliver 10 to 15 drops per milliliter, with 12 drops being a good average.

Measuring Out a Given Mass of Solid Reagent. Solids are not used very often in qualitative analysis procedures, but it is occasionally necessary to add a given amount, e.g., 0.1 g of a solid, to the system under study. Again, this ordinarily is *not* done in what would seem to be the obvious way, namely, by weighing the sample on a balance. A high degree of accuracy is seldom necessary, and an amount as large as 0.15 or even 0.2 g would not be disastrous.

One usually dispenses a small amount of solid from the tip of a spatula. The problem is to determine the space that 0.1 g of solid occupies. To find out, work with a sample of a typical solid, such as $CaCO_3$, powdered limestone, and a triple beam or top-loading balance. Put a small beaker on the balance and weigh it. Then add portions of solid on the end of the spatula, weighing each, until you can add a sample that you are sure weighs 0.1 g ± 0.02 g. If you assume that equal volumes of all solids weigh approximately the same, you will not go too far wrong.

Centrifuging. As we have noted, we invariably use a centrifuge to separate solid precipitates from solutions. Put the test tube containing the precipitate into one of the locations in the centrifuge *and another test tube containing a similar volume of water in a location on the opposite side of the centrifuge.* (You may find such a tube permanently taped in the centrifuge, in which case you should put your test tube in the opposite slot.) Turn on the centrifuge and let it run for at least 30 seconds. Since most precipitates quickly settle to a compact mass in the bottom of the test tube, the time required for centrifuging is short. If your sample is still suspended, centrifuge it again, and if this still does not work, you may find heating the sample in a water bath for a few minutes helps formation of larger crystals of solid, which centrifuge out more easily.

In most cases the supernatant liquid, which we may call the decantate or simply the liquid or the solution, can be poured, or decanted, into another test tube without disturbing the solid. If by chance this cannot be done, and the solid is poured out too, pour everything back into the original test tube, centrifuge, and carefully remove the bulk of the supernatant liquid with a medicine dropper.

Precipitating a Solid. One of the most common reactions in qualitative analysis is the precipitation of substances from solution. To accomplish a precipitation, add the indicated amount of precipitating reagent to the sample solution and stir well with your glass stirring rod. Heat the solution in the water bath if so directed. Since some precipitates form slowly, they must be given time to form completely. It is hard to overstir, but students frequently do not mix reagents thoroughly enough. When the precipitation is believed to be complete, centrifuge out the solid, and before decanting the liquid, add a drop or two of the precipitating reagent, just to make sure that you added enough precipitant the first time. The directions usually specify enough reagent to furnish an excess, but it is good insurance to test, just the same.

Washing a Solid. The liquid decanted from the test tube after centrifuging out a solid does not in general contain any of the solid and may be used directly in a following step. The solid remaining in the tube, however, has residual liquid around it. Since this liquid contains ions that may interfere with further tests on the solid, they must be removed. This is accomplished by diluting the liquid with a wash liquid, often water, which does not interfere with the analysis, dissolve the solid, or precipitate any substances from solution.

To wash the solid, add the indicated amount of wash liquid to the test tube and mix well with your glass stirring rod, dispersing the solid well in the wash liquid. Simply pouring in the wash liquid and pouring it out again is not effective. After mixing thoroughly, centrifuge out the solid and decant the wash liquid, which usually may be discarded. The washing operation, in key separations, is best done twice, because an uncontaminated precipitate can be a joy to behold, and not only that, it tends to give better results.

Heating a Mixture. Many reactions are best carried out when the reagents are hot. We *never*, or almost never, heat a test tube containing a reaction mixture directly in a bunsen flame. It is more convenient and much more considerate of your neighbor to heat the test tube in a water bath. A 150 ml beaker containing about 100 ml of water makes a very adequate water bath.

When you are following a procedure in which you need to heat a solution, keep the water bath hot or boiling by using a small flame to heat the beaker on a piece of asbestos-covered wire gauze. Then the bath is ready whenever required, and you do not have to wait 10 minutes for the water to heat. The test tube, or tubes, can be put directly into the bath, supported against the wall of the beaker. In a water bath you can bring the mixture under study to approximately 100° C without boiling it. If you heat the sample in an open flame, it inevitably bumps out of the tube and onto the laboratory bench, or onto you or someone else. Therefore, *do not heat* test tubes containing liquids over an open flame.

Evaporating a Liquid. In some cases it is necessary to boil down a liquid to a small volume. This may be done to concentrate a species, to remove a volatile reagent, or perhaps because the reaction will proceed readily only in a boiling solution. When this operation is necessary, we perform the boiling in a small 30 ml beaker on a square of asbestos-covered wire gauze and use a small bunsen flame judiciously applied to maintain controlled, gentle boiling. Since the volumes involved usually are of the order of 3 or 4 ml, overheating can easily occur.

If you are supposed to stop the evaporation at a volume of 1 ml or so, make sure that you do not heat to dryness, because you may decompose the sample or render it inert. Since concentrated solutions or slurries tend to bump, it is often helpful to encourage smooth boiling by scratching the bottom of the beaker with a stirring rod as the boiling proceeds. Good judgment in boiling down a sample is important, so use it.

Occasionally the liquid that is evaporated is highly acidic with HCl or

HNO_3, and the boiling causes evolution of noticeable amounts of toxic gases into the laboratory. In some cases the amounts are small enough to be ignored. If you or anyone else notices that bothersome vapors are escaping from your boiling mixture, transfer your operation to the hood. In some laboratories each station has its own small hood; in such a case, carry out all your evaporations under your hood.

Transferring a Solid. Sometimes it is necessary to transfer a solid from the beaker in which it was prepared to a test tube for centrifuging, or from the test tube to the beaker. The amount of solid involved is never large, of the order of 50 mg at most. We always perform such transfers in the presence of a liquid, which serves as a carrier.

When you are ready to make the transfer, stir the solid well into the liquid, forming a slurry, and then, without delay, pour the slurry into the other container. The transfer is not quantitative, but if done properly, you can move 90 per cent of the sample to the other container. Forget about the rest. In general, we *do not* attempt to transfer wet solids with a spatula, because it is not easy, and all too often the spatula reacts with the liquid present, contaminating it.

Handling a Stirring Rod. You will find that your stirring rod is a very useful tool and that you need it in nearly every step in each procedure. The problem is that each time it is used, it gets wet with the solution being treated, and then it must be cleaned before it can be used again. This is a minor problem, but a bothersome one. The best solution we have found is to keep a 250 or 400 ml beaker full of distilled water handy as a storage place for stirring rods. After you have used a rod, swirl it around in the water in the beaker to clean it, and then leave it in the water. Solutes and solids accumulate, but they really amount to tiny traces when diluted in the water. Change the water occasionally to ensure that no significant contamination occurs. We have found this procedure for handling stirring rods to work very well and have had no spurious test results.

Adjusting the pH of a Solution. One of the most important variables controlling chemical reactions is the pH of the solution. Frequently it is necessary to make a basic solution acidic, or vice versa, in order to make a desired reaction take place. For instance, if you are directed to add 6M HCl to an alkaline mixture until it is acidic, you should proceed in the following way. Knowing how the alkaline solution was prepared, make a quick mental calculation of about how much acid is needed—1 drop, 1 ml, or perhaps more. Then add the acid, drop by drop, until you think the pH is about right. Mix well with your stirring rod and then touch the end of the rod to a piece of blue litmus paper on a piece of paper towel or filter paper. If the color does not change, add another drop or two of acid, mix and test again. Frequently the solution changes at the neutral point; a precipitate may dissolve or form, or the color may change. In any event, add enough acid so that, after mixing, the litmus paper turns red when touched with the stirring rod. Similarly, if you are told to make a solution basic with 6M NH_3 or 6M NaOH, add the reagent, drop by drop, until the solution, after being well mixed, turns red litmus blue. Adjustment of pH is not difficult, but it must be done properly if the desired reaction is to occur.

Performing Flame Tests. Some cations are most easily identified in solution by flame tests. For such tests, it is best if the species to be tested for is present either as a solid or in a concentrated solution. You should evaporate the solution nearly to dryness in a beaker, and then add a drop or two of 12M HCl to moisten the residue. A piece of platinum wire sealed into one end of a piece of soft glass tubing should be cleaned—by heating the wire in a bunsen flame until the flame is colorless. Then touch the wire to the solid, picking up some of the solid, and put the wire back into the flame. The color of the flame is indicative of the ions present in the solid.

SAFETY IN THE LABORATORY

It is a common tendency among students, and even among some chemists beginning their professional work, to ignore the fact that working in the laboratory always involves danger. Since reactions proceed smoothly most of the time, with no explosions or fires, we tend to become a bit careless. However, every chemist eventually encounters a chemical accident either with his own reacting system or with that of someone else, and the result of that accident may be trivial or serious, depending upon whether suitable precautions had been taken before it occurred.

The single most important precaution you can take in the laboratory is to protect your eyes by wearing safety glasses. If you wear glasses anyway, they will give you adequate protection. If you do not, you must wear safety glasses at all times, because accidents tend to occur when you least expect them. A tiny drop of hot acid spattering out of a beaker falls harmlessly on a glass lens, but over the years we have seen too many eye injuries that were caused simply because nothing was present to protect the eyes. Therefore, wear your safety glasses and have no regrets.

For the most part, our work in qualitative analysis is safe. Except for chlorates, which we handle very carefully indeed, no potentially explosive substances are used in any reactions. The only inflammable solvents are ethyl alcohol and ether, which are used with caution and in very small amounts. The most hazardous reagent solutions are 18M H_2SO_4, 12M HCl, 17M HNO_3, and 6M NaOH, all of which can cause chemical burns, particularly when hot. You must use care when dispensing these reagents; if you spill one, wash the affected area with water and report the accident to your instructor, even if you think it was unimportant.

GENERAL SUGGESTIONS

There is nothing magic about doing good work in qualitative analysis. The student who is organized, uses good work habits, and has a measure of common sense does better than the student who makes few plans and tends to try to work by just following directions.

Time in the laboratory is spent most effectively if you go over the experiments before the laboratory session. Try to understand what is happening in a given reaction, so that you will know more than just that the solution "turned blue." You may find it very helpful to prepare an outline of the steps you plan to carry out, with enough details so that you can perform them without having to refer again to the text.

It is important in the course of your work to keep good records. Your instructor will require detailed reports of the results of your analyses, and these are much easier to write and much better in quality if your record of what happened is complete and accurate. You should get a laboratory notebook, possibly looseleaf, so that you can store your reports after they have been returned. In any event, have a section in the notebook for each analysis you carry out. There you should jot down what you saw happen in the tests with known solutions, what observations you made on unknown solutions, and, in particular, any anomalous occurrences.

Remember, all your observations in analyzing an unknown should be consistent with the result that you report. If you cannot explain the result of a particular test in terms of the properties of the ions you believe to be present, something is wrong. Either you carried out the test incorrectly or your identification of the unknown is in error.

Good organization in the laboratory also requires a sensible approach to carrying out procedures. Since you are trying to find out what species a sample contains, you certainly do not want to contaminate the sample with residues from dirty test tubes and beakers. Keep those items clean, washing them as a group with soap and water either at the beginning or at the end of each laboratory session. Keep track of precipitates and solutions that you will need later in a procedure, labeling them so that you know what they are. If you need to keep a liquid or solid for a following laboratory period, be especially careful to label it, and put a stopper in the test tube so that the sample does not evaporate or decompose in the air.

As we write these suggestions, they seem so obvious that they should not have to be made, but it is sadly true that many students try to get through their laboratory work with very little effort at organization. They may succeed, but you can be sure that the time required to avoid being organized is much greater than that needed to work sensibly and systematically. Now is a good time to learn to work effectively in a laboratory; you will enjoy it more if you do, and you will do a better job besides.

CHAPTER 3

THE CATIONS

A BRIEF INTRODUCTION

Although the qualitative analysis procedure for a small group of cations allows them to be separated and identified one by one, when one is dealing with a relatively large number of these ions, a more sophisticated approach is required. In the standard qualitative analysis scheme, the cations are removed in small groups from the general mixture. The separation is accomplished by successive precipitations, taking advantage of the similar solubility characteristics of the ions within a group. For the set of cations we are considering, it is convenient to separate the ions into four groups, which are precipitated one after the other from the mixture. The members of each group, and the conditions under which that group is precipitated, are as follows:

Group I Ag^+, Hg_2^{2+}, Pb^{2+} Precipitated as chlorides under
                                                       acidic conditions.

Group II Cu^{2+}, Cd^{2+}, Bi^{3+}, Sn^{2+} Precipitated as sulfides under
   and Sn^{4+}, Hg^{2+}, Sb^{3+} and Sb^{5+},   mildly acidic conditions.
   (Pb^{2+})

Group III Al^{3+}, Zn^{2+}, Cr^{3+}, Fe^{2+} Precipitated as sulfides or
   and Fe^{3+}, Ni^{2+}, Co^{2+}, Mn^{2+} hydroxides under slightly basic
                                                       conditions.

Group IV Ba^{2+}, Ca^{2+}, Mg^{2+}, Na^+,   Remain in solution after precipita-
   K^+, NH_4^+ tion of Group III cations.

After being removed from solution, the cations within a group are further resolved by carrying out complex ion formation or acid-base reactions, which separate the ions into soluble and insoluble fractions. These fractions are then caused to participate in other similar reactions of the types discussed in Chapter 1, until the resolution of the cations is sufficient to allow their identification by one or more specific tests.

This chapter is divided into four sections, one for each of the four groups of cations. Each section begins with a brief description of the characteristic reactions of the cations within a given group. For each cation the description includes all the reactions in which the ion participates in the procedure for analysis, plus other reactions that may serve for the identification of that ion.

29

This discussion is followed by the detailed, step by step procedure for analysis for the cations in that group. We include a set of comments on each procedure, in which we may state the reasons for a given step, describe some of the reactions that occur, and note difficulties and interferences with certain tests. Each section concludes with a flow chart, which summarizes the steps in the procedure.

In the laboratory work on the qualitative analysis of cations, we will study one group at a time, analyzing an "unknown" for the presence of the cations in that group after applying the procedure for analysis to a "known" sample. When the analyses of later groups have been completed, limited unknowns will be assigned, which may contain ions from all the groups up to and including the one being investigated.

SECTION 1

THE PROPERTIES OF THE CATIONS IN GROUP I— Ag^+, Pb^{2+}, Hg_2^{2+}

SILVER, Ag^+

Of the common salts of silver, only the nitrate and fluoride are readily soluble in water. The sulfate, acetate, and chlorate are moderately soluble, ~1 to 10 g/lit, whereas the other salts are all only slightly soluble, ~0.1 g/lit or less. Silver salts are usually white, but the following have the colors noted: bromide (light yellow), iodide and phosphate (yellow), chromate (dark red), and sulfide (black). Silver oxide is dark brown. Silver forms many complex ions and typically has a coordination number of 2, e.g., $Ag(NH_3)_2^+$, $AgCl_2^-$.

CHARACTERISTIC REACTIONS OF THE SILVER ION

1. **Hydrochloric Acid and Soluble Chlorides.**   Silver ion is precipitated from solution as the chloride, AgCl, on addition of HCl or any soluble chloride:

$$Ag^+ + Cl^- \rightleftharpoons AgCl(s)$$

The AgCl is a curdy white precipitate, quite insoluble in water, and also insoluble in dilute acids. The AgCl can be distinguished from other insoluble chlorides by its solubility in 6M NH_3, which occurs by formation of the silver ammonia complex ion:

$$AgCl(s) + 2\ NH_3 \rightleftharpoons Ag(NH_3)_2^+ + Cl^-$$

The reaction is reversed by addition of 6M HNO_3, which reacts with the NH_3 and thereby destroys the complex. The precipitation of AgCl by addition of HNO_3 to its solution in NH_3 provides definitive identification of Ag^+ ion:

$$Ag(NH_3)_2^+ + Cl^- + 2\ H^+ \rightleftharpoons AgCl(s) + 2\ NH_4^+$$

The AgCl is also soluble in cyanide, thiosulfate, and concentrated chloride solutions because it forms silver complex ions.

2. **Sodium Hydroxide.** Addition of this reagent to solutions containing silver ion yields a brown precipitate of Ag_2O, which is not soluble in excess reagent:

$$2\,Ag^+ + 2\,OH^- \rightleftharpoons Ag_2O(s) + H_2O$$

One would expect to get the hydroxide, AgOH, but dehydration to the oxide occurs spontaneously. Silver oxide is soluble in NH_3 as well as in other solutions containing ligands that form silver complex ions.

3. **Ammonia.** As might be expected from the previous discussion, Ag_2O precipitates from solutions of Ag^+ ion on addition of a small amount of 6M NH_3 because of the weakly basic properties of ammonia; it is readily soluble in excess reagent.

4. **Bromides and Iodides.** Addition of solutions containing Br^- or I^- ions produces very insoluble precipitates of AgBr and AgI. These do not dissolve appreciably in 6M NH_3; the bromide is somewhat more soluble than the iodide, and goes into solution in 16M NH_3.

5. **Hydrogen Sulfide and Soluble Sulfides.** These reagents precipitate silver as the highly insoluble black sulfide, Ag_2S, even from 1M acid solutions. The sulfide is insoluble in typical complexing reagents, but it can be brought into solution in warm 6M HNO_3 by oxidation of the sulfide to free sulfur:

$$3\,Ag_2S(s) + 8\,H^+ + 2\,NO_3^- \longrightarrow 6\,Ag^+ + 3\,S(s) + 2\,NO(g) + 4\,H_2O$$

6. **Oxidizing and Reducing Agents.** Reasonably active metals, such as zinc, iron, and copper, readily reduce silver compounds, producing metallic silver. If an insoluble halide of silver is in a mixture with finely divided mercury, the silver may be reduced to the metal. Silver metal is quite soft and ductile; it dissolves readily in 6M HNO_3:

$$3\,Ag(s) + NO_3^- + 4\,H^+ \longrightarrow 3\,Ag^+ + NO + 2\,H_2O$$

LEAD, Pb^{2+}

Most of the salts of lead are insoluble. The nitrate and acetate are the only common salts that are soluble in water; the chloride, bromide, iodide, and thiocyanate are slightly soluble. The three common oxides are all essentially insoluble; PbO, litharge, is yellow, Pb_3O_4 is red, and PbO_2 is dark brown. The oxide PbO is readily soluble in nitric acid; to dissolve PbO_2 or Pb_3O_4 it is necessary to reduce the lead to the Pb(II) state, which can be conveniently done with H_2O_2 in the presence of hot 6M nitric acid:

$$PbO_2(s) + H_2O_2 + 2\,H^+ \longrightarrow Pb^{2+} + 2\,H_2O + O_2(g)$$

CHARACTERISTIC REACTIONS OF THE
LEAD ION

1. **Hydrochloric Acid and Soluble Chlorides.** These reagents precipitate white $PbCl_2$ from lead solutions that are not too dilute. The solubility of lead chloride is about 10 g/lit at 20° C and 33.5 g/lit at 100° C. The solubility of $PbCl_2$ is initially less in chloride solutions than in water, but if the chloride concentration is high, the lead complex ion, $PbCl_4^{2-}$ forms, and the chloride may dissolve:

$$PbCl_2(s) + 2\ Cl^- \rightleftharpoons PbCl_4^{2-}$$

Although Pb^{2+} is considered to be in Group I, the solubility of $PbCl_2$ is too great for one to assume that lead ion is quantitatively precipitated by chloride ion. If the lead ion concentration is great enough to allow detection as a Group I cation, identification at that point is convenient and easy. If the concentration is low, lead ion is detected only in Group II, in which it is quantitatively precipitated as the sulfide.

2. **Dilute Sulfuric Acid and Soluble Sulfates.** Lead sulfate, $PbSO_4$, is much less soluble than the chloride, about 0.04 g/lit, and precipitates in the presence of sulfate ions as a white, finely divided solid. The $PbSO_4$ does not dissolve readily in strong acids, but it goes into solution in a hot concentrated solution of hydroxide or acetate ions. The former complexes the lead as $Pb(OH)_4^{2-}$, whereas the latter produces only slightly dissociated lead acetate:

$$PbSO_4(s) + 4\ OH^- \rightleftharpoons Pb(OH)_4^{2-} + SO_4^{2-}$$
$$PbSO_4(s) + 2\ C_2H_3O_2^- \rightleftharpoons Pb(C_2H_3O_2)_2 + SO_4^{2-}$$

A convenient way to get $PbSO_4$ into solution is to boil it with concentrated Na_2CO_3 solution. Highly insoluble $PbCO_3$ is formed, which can be removed from the system by centrifuging and can then be dissolved in dilute nitric acid:

$$PbSO_4(s) + CO_3^{2-} \rightleftharpoons PbCO_3(s) + SO_4^{2-}$$
$$PbCO_3(s) + 2\ H^+ \longrightarrow Pb^{2+} + CO_2(g) + H_2O$$

3. **Hydrogen Sulfide and Soluble Sulfides.** Black lead sulfide, PbS, precipitates from solutions containing sulfide ion, even in the presence of 1M concentrations of H^+ ion. The sulfide is insoluble in strong acids and bases and in solutions containing complexing ligands; it reacts with hot dilute nitric acid to form free sulfur, nitric oxide, and lead ion.

4. **Potassium Chromate.** Bright yellow lead chromate, $PbCrO_4$, precipitates from lead solutions to which chromate ion is added. The chromate is

insoluble in dilute strong acids, but dissolves in 6M NaOH solution, forming $Pb(OH)_4^{2-}$. The formation of the chromate is frequently used as the confirmatory test for the presence of lead ion in solution.

5. **Sodium Hydroxide.** In the presence of OH^- ion, Pb^{2+} precipitates as white $Pb(OH)_2$, which dissolves readily in excess reagent to form the complex ion, $Pb(OH)_4^{2-}$.

6. **Ammonia.** Lead ion precipitates from ammonia solutions as a white basic salt, e.g., $Pb_2O(NO_3)_2$, which is insoluble in excess reagent.

7. **Oxidizing and Reducing Agents.** Lead metal is a much better reducing agent than is silver; lead readily reduces copper(II) to the metal, but it does not react with either iron(II) or zinc(II) species. The metal is very soft and grayish and has a high density, 11.3 g/cm^3. It is readily soluble in hot dilute HNO_3.

MERCURY(I), Hg_2^{2+}

Mercury forms two ions, Hg_2^{2+} and Hg^{2+}. On solution of mercury metal in hot concentrated nitric acid, the mercury(II), mercuric, Hg^{2+}, ion is formed; if the acid is dilute and cold, and excess mercury is present, the product is the mercury(I), mercurous, Hg_2^{2+}, ion. Since Hg_2Cl_2 is insoluble and $HgCl_2$ is soluble, the mercury(I) ion is in Group I and mercury(II) ion, which forms an extremely insoluble sulfide, is in Group II.

In the presence of species that form insoluble salts or complex ions with Hg(II), Hg(I) salts or solutions typically undergo a disproportionation reaction to form black metallic mercury and the Hg(II) species.

The only reasonably soluble mercury(I) salt is the nitrate, but even this compound requires the presence of excess nitric acid if an insoluble basic salt is to be kept from forming.

CHARACTERISTIC REACTIONS OF THE MERCURY(I) ION

1. **Hydrochloric Acid and Soluble Chlorides.** In the presence of chloride ion, mercury(I) precipitates as white, insoluble Hg_2Cl_2. This substance, sometimes called calomel, is insoluble in all common reagents; it goes into solution in aqua regia. It has a very characteristic reaction with ammonia, in which finely divided mercury (black) and mercury(II) amidochloride (white) are produced in a disproportionation reaction:

$$Hg_2Cl_2(s) + 2 NH_3 \longrightarrow Hg(s) + HgNH_2Cl(s) + NH_4^+ + Cl^-$$

The formation of the black mixture on addition of 6M NH_3 to an insoluble chloride is a definitive test for the presence of mercury(I) ion.

2. **Oxidizing and Reducing Agents.** Reducing ions, particularly Sn^{2+} and Fe^{2+}, readily reduce mercurous ion to the metal. The metal normally exists as a liquid and dissolves only in oxidizing acids, nitric acid being the reagent ordinarily used.

3. **Hydrogen Sulfide.** Mercury(II) sulfide, HgS, and mercury are formed in the presence of solutions containing sulfide ion:

$$Hg_2{}^{2+} + H_2S \longrightarrow HgS(s) + Hg(s) + 2\ H^+$$

4. **Sodium Hydroxide.** Mercury (black) and HgO are formed on addition of hydroxide ion to solutions of mercury(I) salts.

5. **Ammonia.** As noted previously, salts of the mercury(I) ion react with ammonia to form basic Hg(II) compounds and mercury. Neither the amido compound nor the mercury dissolves in excess reagent.

GENERAL DISCUSSION OF PROCEDURE FOR ANALYSIS OF THE GROUP I CATIONS
Ag^+, Pb^{2+}, Hg_2^{2+}

As is apparent from the description of the properties of the cations in Group I, each of the ions forms an insoluble chloride. Since all other common cations form soluble chlorides, it is possible to separate the cations in Group I from the other cations on the basis of chloride precipitation. This precipitation is most conveniently done with dilute HCl solution, because the H^+ ion both prevents possible precipitation of hydroxides and is easily removed at a later stage of the analysis.

Since the properties of the cations in Group I differ appreciably, it is easy to find a procedure for separating and identifying them. In the standard scheme of analysis, lead chloride, $PbCl_2$, is resolved from the other two insoluble chlorides by simply adding water and bringing the system to the boiling point; at $100°$ C, $PbCl_2$ is reasonably soluble, whereas the other two chlorides are not. Centrifuging out the precipitate and decanting the liquid, which contains the Pb^{2+} ion, completes the separation. Lead is confirmed as the yellow chromate by adding K_2CrO_4. The separation of silver and mercury(I) ions is readily accomplished by addition of NH_3 to the $AgCl$–Hg_2Cl_2 precipitate. The silver chloride reacts to form the soluble silver ammonia complex, and the mercury(I) chloride reacts to form a mixture of black metallic mercury and white mercury(II) amidochloride. The black residue remaining after addition of NH_3 is confirmatory for Hg_2^{2+}. Reprecipitation of white AgCl on acidification of the ammonia solution establishes the presence of Ag^+ in the sample.

If both silver and mercury(I) ions are present, with the latter in excess, mercury produced when NH_3 is added to the chloride precipitate tends to reduce $Ag(NH_3)_2^+$, forming silver, with the result that one may get a negative test for silver. Under these conditions the solid that contains Hg, $HgNH_2Cl$, and possibly Ag, is dissolved in aqua regia and the test for silver is made on the solution so obtained; in that solution only mercury(II) species are present, and do not interfere as before.

PROCEDURE FOR ANALYSIS OF GROUP I CATIONS

Unless told otherwise, you may assume that 5 ml of your sample contains the equivalent of about 1 ml of 0.1M solutions of the nitrate salts of one or more of the Group I cations, plus possibly ions from Groups II, III, and IV. Roughly speaking, this amounts to about 10 mg of each cation present. This is a very sufficient amount for good qualitative tests, *as long as you do not lose any component cations* by improperly carrying out or interpreting any step.

Step 1. To 3 ml of your sample in a test tube, add 0.5 ml of 6M HCl. Stir well and centrifuge. Decant the liquid, which may contain ions from groups to be discussed later, into a test tube and save it, if necessary, for further analysis; to make sure precipitation of Group I cations was complete, add 1 drop of 6M HCl to the liquid. Wash the precipitate with 2 ml of water and 3 drops of 6M HCl. Stir well. Centrifuge and discard the wash liquid. Wash the precipitate again with water and HCl; centrifuge and discard the wash.

Step 2. To the precipitate from Step 1, which contains the chlorides of the Group I cations, add about 4 ml of water. Heat in the boiling water bath for at least three minutes, stirring constantly. Centrifuge quickly and decant the liquid, which may contain Pb^{2+}, into a test tube.

Step 3. *Confirmation of the presence of lead.* To the liquid from Step 2 add 2 drops of 6M acetic acid and 3 or 4 drops of 1M K_2CrO_4. The formation of a yellow precipitate of $PbCrO_4$ confirms the presence of lead. Centrifuging out the solid may help with the identification, because the liquid phase is orange.

Step 4. *Confirmation of the presence of mercury.* If lead is present, wash the precipitate from Step 2 with 4 ml of water in the boiling water bath. Centrifuge and test the liquid for Pb^{2+}. Continue the washings until no positive reaction to the lead test is obtained. To the washed precipitate add 2 ml of 6M NH_3 and stir well. A black or dark gray precipitate establishes the presence of the mercury(I) ion. Centrifuge and decant the liquid, which may contain $Ag(NH_3)_2^+$, into a test tube.

Step 5. *Confirmation of the presence of silver.* To the liquid from Step 4 add 3 ml of 6M HNO_3. Check with litmus to see that the solution is acidic. A white precipitate of AgCl confirms the presence of silver.

Step 6. *Alternative confirmation of the presence of silver.* If the test for silver ion was inconclusive, and mercury was present, add 1 ml of 12M HCl and 0.5 ml of 6M HNO_3 to the precipitate from Step 4. Heat in the water bath until solution is essentially complete. Pour the liquid into a 30 ml beaker and boil it gently for a minute. Add 3 ml of water. If a white precipitate forms, it is probably AgCl. Centrifuge and discard the liquid. To the precipitate, add 0.5 ml of 6M NH_3; with stirring, the precipitate should dissolve. Add 1 ml of 6M HNO_3; if silver is present, a precipitate of AgCl forms.

COMMENTS ON PROCEDURE FOR ANALYSIS OF GROUP I CATIONS

Step 1. In this step the cations of Group I are precipitated as their chlorides. Both silver and lead ions can form chloride complexes in solutions when the chloride ion concentrations are high. In our procedure, $[Cl^-]$ is about 1M, which decreases the salt solubilities by the common ion effect, but is not great enough to cause appreciable amounts of the complex ions to form. It is necessary to precipitate all of the Group I cations at this point; although the amount of HCl added should be ample, a check for complete precipitation should be made.

Step 2. The $PbCl_2$ is sufficiently soluble in hot water to allow its separation from the other chlorides by simply heating the precipitate, with mixing, in water. The centrifuging should be done quickly to avoid reprecipitation of $PbCl_2$ on cooling.

Step 3. We acidify the liquid to minimize precipitation of other chromates from residual amounts of ions in other groups. The liquid is orange because of the conversion of CrO_4^{2-} to $Cr_2O_7^{2-}$ in the acid medium.

Step 4. If lead chloride is not completely removed, it is converted to a white, basic, insoluble, salt on addition of NH_3. This could cause confusion, but should not interfere with later identifications. The AgCl dissolves readily in 6M NH_3, with formation of the silver ammonia complex ion. If Hg_2Cl_2 is present, it reacts with NH_3, forming black Hg and white insoluble $HgNH_2Cl$. The mixture is black or dark gray if mercury(I) ion is present.

Step 5. The silver ammonia complex ion is destroyed by acid, and the released silver ion precipitates with the chloride ion in solution. The formation of white AgCl is definitive evidence for the presence of silver.

Step 6. If both mercury(I) ion and silver ion are present, the $Ag(NH_3)_2^+$ and mercury, both present in Step 4, tend to undergo an oxidation-reduction reaction:

$$2\,Ag(NH_3)_2^+ + Hg(s) + Cl^- \longrightarrow 2\,Ag(s) + HgNH_2Cl(s) + NH_4^+ + 2\,NH_3$$

If sufficient mercury is present, nearly all the silver may be reduced and a very doubtful test for Ag^+ obtained in Step 5. If it appears that this is the case, carrying out Step 6 should be helpful. On being dissolved in aqua regia, the mercury exists as $HgCl_4^{2-}$ and the silver as $AgCl_2^-$. Adding water to the strongly acid solution reprecipitates white AgCl, which can then be separated from the $HgCl_4^{2-}$ by centrifuging. The solution of AgCl in NH_3 and reprecipitation on addition of HNO_3 is definitive confirmation of the presence of silver.

Outline of Procedure for Analysis of Group I Cations

Ions possibly present:

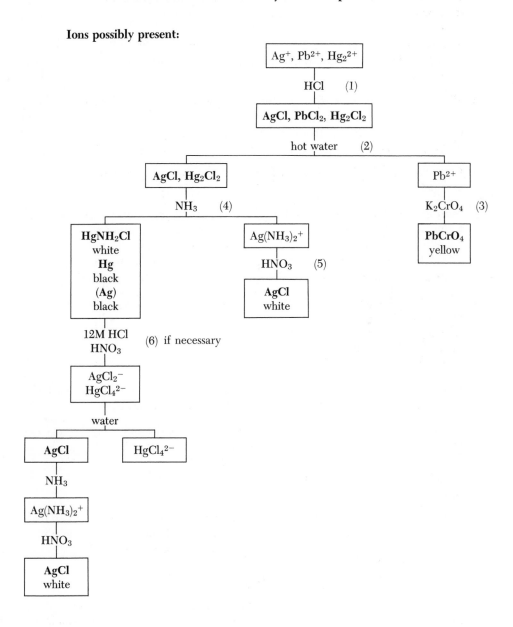

PROBLEMS

1. Write balanced net ionic equations to explain the following observations:
   a. A precipitate forms when solutions of $Pb(NO_3)_2$ and NaCl are mixed.
   b. Silver chloride dissolves when treated with aqueous NH_3.
   c. When the solution formed in (b) is treated with nitric acid, a precipitate forms.
   d. A precipitate forms when solutions of mercury(I) nitrate and hydrochloric acid are mixed.
   e. When a small amount of OH^- is added to a solution of $Pb(NO_3)_2$, a precipitate forms. This precipitate disappears as more OH^- is added. (Two reactions are involved).

2. Write balanced net ionic equations for the following oxidation-reduction reactions:
   a. Silver sulfide is treated with 6M HNO_3.
   b. Mercury(I) chloride is treated with ammonia.
   c. A strip of copper is added to a solution of silver nitrate.

3. Give the formula of a reagent that brings the following solids into solution:
   a. AgCl   b. $Pb(NO_3)_2$ c. $PbCl_2$ d. Ag

4. Describe a simple test that would enable you to distinguish between the following:
   a. AgCl(s) and $AgNO_3$(s)   e. Ag^+ and Hg_2^{2+} (in solution)
   b. $Pb(NO_3)_2$(s) and $NaNO_3$(s)   f. Hg_2^{2+} and Pb^{2+} (in solution)
   c. AgCl(s) and $PbCl_2$(s)   g. NO_3^- and Cl^- (in solution)
   d. $PbCO_3$(s) and $PbSO_4$(s)   h. Cl^- and I^- (in solution)

5. What, if anything, can be deduced about the identity of a Group I unknown from the following observations?
   a. A precipitate is formed in Step 1.
   b. In Step 3, the solution is a bright orange.
   c. A precipitate forms in Step 4.

6. Suppose you are working with a Group I unknown containing only Ag^+ and Pb^{2+}. State precisely what will be observed at each step of the analysis.

7. A student, in analyzing a Group I unknown, makes the following "errors" in procedure. What effect, if any, will they have on the results?
   a. In Step 1, he uses nitric rather than hydrochloric acid.
   b. In Step 5, he uses hydrochloric rather than nitric acid.
   c. In Step 4, he uses NaOH instead of NH_3.

8. A student is told that he is to analyze an unknown that can contain no more than two cations. He is asked to devise an abbreviated scheme of analysis applying to only these two ions and including no extra steps. Outline such a scheme, assuming that the ions are the following:
   a. Ag^+ and Pb^{2+} b. Ag^+ and Hg_2^{2+}

9. A student, in analyzing a Group I unknown, makes the following observations. Indicate which ions are definitely present, which are absent, and which are questionable, i.e., might be present or absent. *Note that in this problem and others like it throughout this text, confirmatory tests are ordinarily not included.*

The precipitate formed when HCl is added to the unknown is partially soluble in hot water. The precipitate remaining is completely soluble in 6M NH_3.

10. A student is given a solid unknown that may contain one or more of the following:

$$AgCl, \; PbSO_4, \; Hg_2Cl_2, \; AgNO_3$$

The solid is first stirred with water. A precipitate forms when the solution is treated with HCl. The solid remaining from the first step is treated with 6M NH_3. The solid turns black; the solution formed with NH_3 gives no precipitate when treated with HNO_3.

On the basis of these observations, state which solids are definitely present, which are absent, and which are questionable.

101. Write balanced net ionic equations for the reactions that occur when
    a. Solutions of $Hg_2(NO_3)_2$ and NaCl are mixed.
    b. A solution formed by dissolving AgBr in NH_3 is acidified.
    c. Solutions of $AgNO_3$ and $Ca(OH)_2$ are mixed.
    d. A hot, aqueous solution of lead chloride is cooled.
    e. Successively greater concentrations of Cl^- are added to a concentrated solution of $AgNO_3$. (Two reactions occur).

102. Write balanced net ionic equations to describe the following oxidation-reduction reactions:
    a. Silver metal goes into solution when treated with HNO_3.
    b. Hydrogen sulfide is bubbled into a water solution of a mercury(I) salt.
    c. Lead ion is oxidized to $PbO_2(s)$ when treated with ClO_3^- in basic solution. (Assume reduction of ClO_3^- to Cl^-.)

103. Name a reagent that will bring the following into solution:
    a. $AgNO_3$ b. AgBr   c. Ag_2S d. $PbSO_4$

104. Describe a simple test that would enable you to distinguish between the following:
    a. AgCl(s) and AgI(s)   e. Ag^+ and Pb^{2+}
    b. $Hg_2(NO_3)_2$ and HNO_3 f. Ag^+ and $Ag(NH_3)_2^+$
    c. $Hg_2Cl_2(s)$ and $PbCl_2(s)$ g. Pb^{2+} and $Pb(OH)_4^{2-}$
    d. $Pb(NO_3)_2(s)$ and $PbSO_4(s)$ h. Hg_2^{2+} and Hg^{2+}

105. What, if anything, can be deduced about the identity of a Group I unknown from the following observations?
    a. The precipitate formed in Step 1 is white.
    b. Precipitates are formed in Steps 1 and 5.
    c. In Step 2, part of the precipitate dissolves; no precipitate forms when CrO_4^{2-} is added.
    d. A precipitate forms in Step 1 but not in Steps 3, 4, and 5.

106. A Group I unknown contains only Pb^{2+} and Hg_2^{2+}. State precisely what will be observed in each step of the analysis.

107. Describe what effect, if any, the following "errors" in procedure have on the results of analysis of a Group I cation.

    a. In Step 2, the solution is allowed to cool to room temperature before being centrifuged.

    b. In Step 4, the precipitate is not washed before adding ammonia.

    c. In Step 5, NH_3 rather than HNO_3 is added.

108. Devise an abbreviated scheme of analysis, containing no extra steps, for an unknown that contains no cations other than the following:

    a. Hg_2^{2+} and Pb^{2+} b. $Ag(NH_3)_2^+$ and Pb^{2+}

109. A student, in analyzing a Group I unknown, makes the following observations. Indicate which ions are definitely present, which are absent, and which are questionable.

    The precipitate formed on addition of HCl is completely insoluble in hot water; the residue gives a precipitate on treatment with ammonia.

110. A solid may contain any of the following:

$$PbCl_2, \ Hg_2(NO_3)_2, \ Ag_2S, \ PbCO_3$$

Describe a scheme of analysis that would enable you to determine which of these solids are present in the unknown.

LABORATORY ASSIGNMENTS

Perform one or more of the following, as directed by your instructor.

1. Make up a sample containing about 1 ml of 0.1M solutions of the nitrate salts of each of the cations in Group I. Go through the standard procedure for analysis of Group I cations, comparing your observations with those that are given. Then obtain a Group I unknown from your instructor and analyze it to determine the possible presence of Ag^+, Pb^{2+}, and Hg_2^{2+}. On a Group I flow chart indicate your observations and conclusions on the unknown and turn the completed chart in to your instructor.

2. You will be given an unknown solution that contains only one of the Group I cations and no other cations. Develop the simplest possible procedure you can think of to determine which cation is present. Test your procedure to see that it works, and then use it to analyze an unknown that your instructor furnishes you. Draw a flow chart showing your procedure and the observations to be expected at each step with any one of the possible cations. In another color indicate your observations with the unknown and your conclusions. Your grade depends on your having a workable, brief procedure and a proper analysis. (The simplest approach under the condition that only one Group I cation may be present bears little resemblance to the general scheme for analysis of Group I cations.)

3. If a sample may contain only the cations in Group I, it is possible to analyze it by several procedures that are quite different from the one given in this text. Develop a scheme for analysis of such a sample, starting with the addition of 6M NaOH in excess. Draw the complete flow chart for your procedure, indicating reagents to be added at each step and the formulas of all species present during

the course of the analysis. Test your procedure with a Group I known, and then use your method to analyze a sample of unknown. In another color indicate on the flow chart your observations on the unknown and your conclusions regarding its composition.

NOTE: In this and all succeeding laboratory assignments, the first assigned problem involves your acquiring familiarity with a standard procedure for analysis. Succeeding problems have as their purpose your developing and using your own schemes for analyzing particular unknown mixtures. In setting up your procedures, you may use steps from the standard approaches, but you should also examine the characteristic reactions of the individual ions to see if some of them might be profitably used in your scheme. In all probability, the best method of analysis for any given limited mixture of ions will be in part standard and in part based on ion properties that were not made use of in the standard approach.

THE PROPERTIES OF THE CATIONS IN GROUP II— Cu^{2+}, Bi^{3+}, Hg^{2+}, Cd^{2+}, (Pb^{2+}), Sn^{2+} and Sn^{4+}, Sb^{3+} and Sb^{5+}

COPPER, Cu^{2+}

In its compounds copper can exist in either the $+1$ or the $+2$ state. Copper(I), cuprous, compounds are relatively rarely encountered; they are typically insoluble in water, and except for the sulfide, which is black, and the oxide, which is red, they are all colorless. The solution chemistry of the copper ion, at least for the most part, is that of the copper(II), cupric, Cu^{2+}, ion.

Most copper(II) salts exist as hydrates in the solid state and are either green or blue. The hydrated ion, $Cu(H_2O)_4^{2+}$, is blue; consequently, except in the presence of strong complexing agents, solutions containing copper(II) ion are blue. Copper(II) salts are for the most part insoluble, with the important exceptions being the sulfate, nitrate, chloride, bromide, and acetate. Copper(II) oxide, CuO, is black and insoluble in water, but dissolves in solutions of strong acid; CuO is a typical transition metal oxide. Because the copper(II) ion forms many complexes, most of which are colored, its chemistry is quite distinctive and interesting.

CHARACTERISTIC REACTIONS OF THE COPPER(II) ION

1. **Hydrogen Sulfide.** Even in moderately acidic solutions hydrogen sulfide precipitates black copper(II) sulfide, CuS, from solutions of copper ion:

$$Cu^{2+} + H_2S \longrightarrow CuS(s) + 2 H^+$$

The CuS is insoluble in dilute acids and bases, and, unlike many sulfides, does not dissolve appreciably in hot NaOH solution containing sulfide ion. It does, however, go into solution readily in hot 6M HNO_3:

$$3 CuS(s) + 2 NO_3^- + 8 H^+ \longrightarrow 3 Cu^{2+} + 2 NO + 3 S(s) + 4 H_2O$$

2. **Sodium Hydroxide.**   This reagent precipitates light blue, flocculent copper(II) hydroxide, $Cu(OH)_2$; in very concentrated solutions of NaOH one can dissolve $Cu(OH)_2$, forming a blue complex ion, presumably $Cu(OH)_4^{2-}$.

3. **Ammonia.**   Addition of NH_3 solution initially precipitates a light blue basic salt or copper(II) hydroxide, but in excess reagent these substances readily dissolve because of formation of the very characteristic dark blue copper(II) ammonia complex ion:

$$Cu(OH)_2(s) + 4\,NH_3 \rightleftharpoons Cu(NH_3)_4^{2+} + 2\,OH^-$$

The $Cu(NH_3)_4^{2+}$ complex ion is quite stable; in fact, it is so stable that even very significant amounts of 6M NaOH solution do not precipitate the hydroxide from a solution of copper ion in 6M NH_3.

4. **Potassium Ferrocyanide.**   In the presence of this reagent, Cu^{2+} precipitates as red-brown copper(II) ferrocyanide, $Cu_2[Fe(CN)_6]$:

$$2\,Cu^{2+} + [Fe(CN)_6]^{4-} \longrightarrow Cu_2[Fe(CN)_6](s)$$

This reaction is probably the most sensitive test for copper(II), giving a red coloration even in extremely dilute solutions. The precipitate is soluble in NH_3 solution.

5. **Potassium Iodide.**   When this reagent is added to a solution containing copper ion, an oxidation-reduction reaction occurs; brown I_2 is formed along with a white precipitate of copper(I) iodide.

$$2\,Cu^{2+} + 4\,I^- \longrightarrow 2\,CuI(s) + I_2$$

6. **Oxidizing and Reducing Agents.**   Cupric ion is readily reduced to the metal by more active metals, such as iron or zinc. An iron nail in a solution of cupric ions becomes coated with a red deposit of metallic copper. Copper metal dissolves readily in hot 6M HNO_3:

$$3\,Cu(s) + 2\,NO_3^- + 8\,H^+ \longrightarrow 3\,Cu^{2+} + 2\,NO(g) + 4\,H_2O$$

BISMUTH, Bi $^{3+}$

Bismuth ion in solution is ordinarily in the $+3$ state, although compounds are known in which this element is in the $+5$ state. Bismuth salts are nearly all colorless; on addition to water they typically react to form insoluble basic salts:

$$Bi^{3+} + Cl^- + H_2O \rightleftharpoons BiOCl(s) + 2\,H^+$$

The basic salt, bismuth oxychloride, dissolves in solutions of strong acids by the reverse of the preceding reaction; it may be assumed that in any solution containing appreciable amounts of Bi^{3+} the concentration of H^+ ion is of the order of 1M or greater.

CHARACTERISTIC REACTIONS OF THE BISMUTH ION

1. **Hydrogen Sulfide.** Like all of the cations in Group II, bismuth ion precipitates as the sulfide in moderately acidic H_2S solutions. Bismuth trisulfide, Bi_2S_3, is dark brown and insoluble in cold dilute acids and bases. It goes into solution in hot 6M HNO_3 and in 12M HCl. It is not soluble in sodium hydroxide solution.

2. **Water.** If a solution of a bismuth salt is added to a large volume of water, the bismuth precipitates as a white basic salt. In the absence of salts of antimony, which have the same property, this reaction can serve as a definitive test for the presence of bismuth ion:

$$Bi^{3+} + Cl^- + H_2O \rightleftharpoons BiOCl(s) + 2\,H^+$$

3. **Sodium Hydroxide and Ammonia.** Solutions of these bases precipitate white bismuth hydroxide, $Bi(OH)_3$, which is insoluble in excess of either reagent. The hydroxide dissolves in moderately strong acid.

4. **Oxidizing and Reducing Agents.** Bismuth ions can be reduced to the metal by a solution of divalent tin, Sn(II). The simplest equation for the reaction is:

$$2\,Bi^{3+} + 3\,Sn^{2+} \longrightarrow 2\,Bi(s) + 3\,Sn^{4+}$$

The reaction occurs, however, only in basic solution, in which bismuth(III) exists as the hydroxide and tin(II) and tin(IV) are present as the hydroxo complexes, $Sn(OH)_4^{2-}$ and $Sn(OH)_6^{2-}$:

$$2\,Bi(OH)_3(s) + 3\,Sn(OH)_4^{2-} \longrightarrow 2\,Bi(s) + 3\,Sn(OH)_6^{2-}$$

The formation of black metallic bismuth in this reaction is often used as a test for bismuth. Bismuth metal is hard and brittle and has a reddish cast. Bismuth dissolves in hot 6M HNO_3 and in hot 18M H_2SO_4.

MERCURY(II), Hg^{2+}

Mercury(II) compounds are typically colorless and are notorious poisons. The soluble salts include the chloride, bromide, cyanide, and acetate. The nitrate tends to react with water to form basic salts, but it dissolves readily in dilute acids. Mercury(II) salts in solution are often only very slightly ionized and form complexes quite readily. The oxide, HgO, exists in both yellow and red forms.

CHARACTERISTIC REACTIONS OF THE MERCURY(II) ION

1. **Hydrogen Sulfide.** In the presence of H_2S, Hg^{2+} ion precipitates as black mercuric sulfide, HgS, the most insoluble of all the sulfides. The HgS is unaffected by hot 6M HNO_3 or hot 12M HCl, but it goes into solution by formation of the complex HgS_2^{2-} ion in hot dilute NaOH solution containing sulfide ion. Mercuric sulfide is also soluble in aqua regia:

$$3\ HgS(s) + 12\ Cl^- + 2\ NO_3^- + 8\ H^+ \longrightarrow$$
$$3\ HgCl_4^{2-} + 2\ NO + 3\ S(s) + 4\ H_2O$$

2. **Sodium Hydroxide.** With this reagent one obtains a precipitate of yellow HgO, which does not dissolve in excess OH^- ion but is soluble in acids.

3. **Potassium Iodide.** The Hg^{2+} ion reacts with iodide ions to form a beautiful red precipitate of HgI_2, which dissolves rather easily in excess reagent by forming the colorless HgI_4^{2-} complex ion.

4. **Ammonia.** Solutions of NH_3 precipitate white, very slightly soluble, basic amido salts, e.g., $HgNH_2Cl$, which are not soluble in excess reagent. These salts dissolve in acids:

$$HgNH_2Cl(s) + 2\ H^+ + Cl^- \rightleftharpoons HgCl_2 + NH_4^+$$

5. **Reducing Agents.** Mercury(II) compounds in solution are readily reduced to either mercury(I) or metallic mercury. The formation of a deposit of bright mercury on a clean piece of copper is a common confirmatory test for the presence of mercury ions in solution. Tin(II) chloride, $SnCl_2$, in acid solution reduces mercury(II) compounds to white Hg_2Cl_2 and metallic mercury.

CADMIUM, Cd^{2+}

Cadmium ordinarily exists in its compounds only in the $+2$ state. Its salts are usually colorless. The soluble salts include the nitrate, chloride, bromide, iodide, sulfate and acetate. Cadmium ion forms many complexes, including those with ammonia, halide ions, and cyanide ion. Yellow-brown CdO is the only oxide. A typical basic oxide, it is insoluble in water and dissolves in acid solution.

CHARACTERISTIC REACTIONS OF THE CADMIUM ION

1. **Hydrogen Sulfide.** A yellow precipitate of CdS is formed in mildly acidic H_2S solutions containing cadmium ion. This sulfide is more soluble than are some of the others in Group II and dissolves in concentrated chloride, bromide, and iodide solutions by forming a complex ion. The sulfide is also soluble in hot 6M HNO_3 and hot 6M H_2SO_4:

$$CdS(s) + 2\,H^+ \rightleftharpoons Cd^{2+} + H_2S$$

The CdS does not dissolve in NaOH solution containing sulfide ion.

2. **Sodium Hydroxide.** White cadmium hydroxide, $Cd(OH)_2$, precipitates in the presence of this reagent and does not dissolve in excess reagent.

3. **Ammonia.** This reagent initially produces $Cd(OH)_2$, which dissolves in excess NH_3 because of formation of the $Cd(NH_3)_4{}^{2+}$ complex ion. A white basic salt, essentially insoluble in ammonia, precipitates readily on addition of 6M NaOH to a solution containing $Cd(NH_3)_4{}^{2+}$.

4. **Oxidizing and Reducing Agents.** Cadmium metal is a relatively good reducing agent, ranking just below iron in this regard. The usual solvent is hot 6M HNO_3, but cadmium also dissolves slowly in dilute HCl and in dilute H_2SO_4.

TIN, Sn^{2+} AND Sn^{4+}

Tin in its compounds and in solution exists in either the $+2$, stannous, or the $+4$, stannic, state. The Sn(II) species, particularly in basic solution, are good reducing agents, and, as we have noted earlier, can reduce Bi(III) and Hg(II) to the respective metals.

Tin in either oxidation state forms many complex ions; in solution tin is usually present as a complex anion. Typical Sn(IV) species include $SnCl_6^{2-}$ and $Sn(OH)_6^{2-}$; in basic solution Sn(II) exists as $Sn(OH)_4^{2-}$. The chlorides, $SnCl_2 \cdot 2 H_2O$ and $SnCl_4 \cdot 5 H_2O$, both hydrolyze in water, forming acidic solutions, and are soluble in acidic and in basic media. Tin(IV) oxide, SnO_2, is inert to most reagents; the best solvent is 18M H_2SO_4 or hot NaOH. Most tin compounds are colorless.

CHARACTERISTIC REACTIONS OF
TIN(II) AND TIN(IV) IONS

1. **Hydrogen Sulfide.** In mildly acidic H_2S solutions precipitates of stannous sulfide, SnS (brown), and stannic sulfide, SnS_2 (light yellow), are produced. The SnS_2 dissolves in dilute NaOH containing S^{2-} ion and in dilute HCl solution, forming complex sulfide and chloride ions, respectively. Sn(IV) forms a very stable oxalate complex, which prevents precipitation of SnS_2 by H_2S when oxalate ion is present. Stannous sulfide is not soluble in NaOH solution but dissolves in 12M HCl, forming the $SnCl_4^{2-}$ complex.

2. **Sodium Hydroxide.** This reagent produces white precipitates of $Sn(OH)_2$ or $Sn(OH)_4$ from solutions of Sn(II) or Sn(IV). These substances both dissolve readily in excess reagent, yielding $Sn(OH)_4^{2-}$ and $Sn(OH)_6^{2-}$ complex ions, respectively. (There is some doubt about the formula of the tin(II) hydroxide complex ion; it may be $Sn(OH)_3^-$.) The $Sn(OH)_4$ forms $Sn(OH)_6^{2-}$ so readily that it is soluble in Na_2CO_3 solution in which the concentration of OH^- is relatively low.

3. **Ammonia.** White hydroxides are formed in NH_3; they are not soluble in excess reagent.

4. **Reducing agents.** In HCl solution metallic iron or aluminum reduces Sn(IV) to Sn(II):

$$Fe(s) + SnCl_6^{2-} \longrightarrow Fe^{2+} + SnCl_4^{2-} + 2 Cl^-$$

Stannous ion in acid solution reduces $HgCl_2$ to form a white precipitate of Hg_2Cl_2, which may turn gray as the reduction proceeds all the way to metallic mercury. In basic solution Sn(II) reduces bismuth ion to the metal. (See confirmatory test for bismuth.)

Tin is a relatively soft, ductile metal. It dissolves in hot 12M HCl, and in 18 M H_2SO_4 with the evolution of SO_2:

$$Sn(s) + 4 H_2SO_4 \longrightarrow Sn^{4+} + 2 SO_2(g) + 2 SO_4^{2-} + 4 H_2O$$

In concentrated HNO_3, tin reacts to form insoluble metastannic acid, $SnO_2 \cdot H_2O$:

$$Sn(s) + 4\,NO_3^- + 4\,H^+ \longrightarrow SnO_2 \cdot H_2O(s) + 4\,NO_2(g) + H_2O$$

ANTIMONY, Sb^{3+} AND Sb^{5+}

Antimony salts, like those of tin, are typically insoluble in water. In solution, antimony, also like tin, is ordinarily present as a complex anion. The most common salt of antimony is $SbCl_3$, which dissolves readily in moderately concentrated HCl, forming $SbCl_6^{3-}$, and in basic solutions, forming $Sb(OH)_4^-$ complex ion. The Sb(III) salts are more common than are those of Sb(V); the latter can be produced from Sb(III) by using a strong oxidizing agent, such as nitric acid.

CHARACTERISTIC REACTIONS OF Sb(III) AND Sb(V) IONS

1. **Hydrogen Sulfide.** Under moderately acid conditions, H_2S precipitates orange-red Sb_2S_3 and Sb_2S_5 from solutions of Sb(III) and Sb(V), respectively. These sulfides are soluble in hot dilute NaOH containing S^{2-} ion and in hot 12M HCl because of formation of very stable sulfide and chloride complex ions.

2. **Water.** Solutions of antimony(III) chloride in HCl form the basic white insoluble salt, SbOCl, when added to an excess of water. The reaction is very similar to that shown by bismuth salts in solution. If necessary, one can distinguish between the two precipitates on the basis that the antimony salt is soluble in either tartaric acid or NaOH solution by complex ion formation, whereas the bismuth salt is not:

$$2\,SbOCl(s) + 3\,C_4H_4O_6^{2-} + 4\,H^+ \rightleftharpoons Sb_2(C_4H_4O_6)_3 + 2H_2O + 2Cl^-$$

3. **Sodium Hydroxide.** White hydroxides are precipitated by this reagent. They are amphoteric and dissolve easily in excess hydroxide or in strong HCl.

4. **Ammonia.** This reagent precipitates a basic salt from solutions of antimony in either oxidation state; the salt does not dissolve in excess NH_3 but is soluble in strongly acidic and in strongly basic systems.

5. Oxidizing and Reducing Agents. When metallic aluminum or iron is immersed in an HCl solution of an antimony salt, hydrogen is evolved and antimony is reduced to the metallic state and is deposited as black particles. Bismuth and lead compounds in solution also form black metallic deposits under these conditions. Antimony metal is white and crystalline. It is soluble only in aqua regia; in nitric acid by itself one obtains insoluble oxides. In 6M HNO_3 the reaction is:

$$2\ Sb(s) + 2\ NO_3^- + 2\ H^+ \longrightarrow Sb_2O_3(s) + 2\ NO(g) + H_2O$$

GENERAL DISCUSSION OF PROCEDURE FOR ANALYSIS OF THE GROUP II CATIONS Cu^{2+}, Bi^{3+}, Hg^{2+}, Cd^{2+}, Sn^{2+} and Sn^{4+}, Sb^{3+} and Sb^{5+}, (Pb^{2+})

Group II contains nine possible cations, which tends to make the analysis a good deal more difficult than that for Group I cations. Successful analysis for Group II cations requires both careful procedure and some knowledge of what is happening at each step and what is expected. Students who proceed properly have little difficulty in achieving good results, but those who try to rely on following the "cookbook" will probably get into trouble.

From the description of the properties of the cations in Group II, one can derive many of the steps in the analysis. These cations all precipitate as sulfides in strongly acid solution, which distinguishes them from other cations, and is, indeed, the prime characteristic of the ions in Group II. Following their precipitation, the sulfides can be easily separated into two subgroups on the basis that CuS, Bi_2S_3, CdS, and PbS do not dissolve in highly alkaline solutions containing sulfide ions, whereas SnS_2, Sb_2S_5, and HgS, because of their tendency to form complex sulfide ions, go into solution under those conditions.

The sulfides in the copper subgroup are then dissolved in nitric acid and treated with ammonia; this complexes both $Cu(II)$ and $Cd(II)$ but not $Bi(III)$ or $Pb(II)$, which precipitate as the hydroxide and as a basic salt, respectively. The presence of copper is established by observation of the blue copper ammonia complex ion and that of cadmium by precipitation, under proper conditions, of yellow CdS. Bismuth ion is separated from lead by treatment of the hydroxides with HCl, which dissolves $Bi(OH)_3$ but converts the lead salt to insoluble $PbCl_2$. Bismuth is identified by precipitation of BiOCl or by reduction to the metal with tin(II) under basic conditions. Lead is confirmed by precipitation as $PbSO_4$ after solution of $PbCl_2$ in acetic acid.

The members of the tin subgroup are reprecipitated as the sulfides, and HgS is separated out by taking advantage of its insolubility in concentrated hydrochloric acid; in that acid, SnS_2 and Sb_2S_5 dissolve, forming $SnCl_6^{2-}$ and $SbCl_6^{3-}$, respectively. Mercuric sulfide is dissolved in aqua regia, and the presence of mercury is established by characteristic reactions with copper metal, stannous chloride, or iodide ion. The solution containing tin and antimony complex ions is divided. One half is tested for the presence of tin by reduction of antimony to the metal and tin to the $+2$ state; the characteristic reaction between $HgCl_2$ and $Sn(II)$ is then carried out. Antimony is observable in several parts of the procedure, but it is confirmed finally by its precipitation as the sulfide from the other half of the solution under conditions in which tin sulfide cannot form.

PROCEDURE FOR ANALYSIS OF GROUP II CATIONS

Unless directed otherwise, you may assume that a 10 ml sample contains about 1 ml of 0.1M solutions of the nitrate or chloride salts of one or more of the Group II cations; your instructor will tell you whether Pb^{2+} is to be considered in the analysis. If you are working with a general unknown, your sample is the HCl solution decanted from the Group I precipitate.

Step 1. Pour 5 ml of your Group II sample or general unknown into a 30 ml beaker. Add 0.5 ml of 6M HNO_3, and boil the solution carefully down to a volume of about 1 ml.

Step 2. Add 2 ml of water to the liquid in the beaker, and swirl it around to dissolve any salt that may have crystallized. Pour the resulting solution into a test tube, and add 6M HCl, drop by drop, until the pH becomes 0.5. (The way to accomplish this is discussed in the Comments on Procedure at the end of this section.) When the pH has been properly established, add 1 ml of 1M thioacetamide to the solution and stir well.

Heat the test tube in the boiling water bath for at least five minutes. If any Group II ions are present, a precipitate forms; typically its color is initially light and gradually darkens, finally becoming black. Continue to heat the test tube for at least two minutes after the color stops changing. Centrifuge out the precipitate, and decant the solution into a test tube. Wash the precipitate with 1 ml of 1M NH_4Cl and 1 ml of water.

Check the pH of the decanted solution; if it is too low, too acid, add 2M $NaC_2H_3O_2$ drop by drop, until the pH is again 0.5. A brown or yellow precipitate may then form, indicating that not all of the Group II cations precipitated the first time. In any event, add 0.5 ml of 1M thioacetamide to the decanted solution at pH 0.5, and heat for three minutes in the boiling water bath. Centrifuge out any precipitate, and decant the liquid, testing it once again for proper pH and for complete precipitation. All the Group II cations must be precipitated; since CdS is relatively soluble, conditions must be just right for it to precipitate properly.

When you are sure that all Group II cations have been precipitated, save the liquid for further analysis if it may contain ions from groups to be studied later. Wash each batch of precipitate with 1 ml of 1M NH_4Cl and 1 ml of water, mix well, and combine all the batches in one test tube. Centrifuge, and discard the wash liquid. The precipitate contains all of the Group II sulfides. Wash the precipitate once again with 1 ml of 1M NH_4Cl and 1 ml of water. Centrifuge, and discard the wash.

Step 3. To the precipitate from Step 2 add 1 ml of water, 1 ml of 1M thioacetamide, and 1 ml of 6M NaOH. Stir, and transfer the slurry completely into a 30 ml beaker. Boil gently for five minutes, with stirring. Any residue is black or yellow. Centrifuge and decant the solution into a test tube. The solution (yellow) may contain SbS_4^{3-}, SnS_3^{2-}, and HgS_2^{2-}; the precipitate may include CuS, Bi_2S_3, CdS, and PbS. Wash the precipitate with 1 ml of 6M NaOH and 2 ml of water. Stir well and centrifuge. Discard the wash.

Step 4. To the precipitate from Step 3 add 1 ml of water and 1 ml of 6M HNO_3. Heat in the boiling water bath until the reaction is complete. The precipitate gradually lightens in color as the reaction proceeds. If a solid coagulates and floats on top, try to immerse it with your stirring rod. The final residue should be tan or light brown; do not stop heating while a black residue is still present. The residue may contain a little HgS, but it is mostly sulfur. Centrifuge and discard the residue. The solution may contain Cu^{2+}, Bi^{3+}, Cd^{2+}, and possibly some Pb^{2+} that was not recovered in Group I.

Step 5. To the solution from Step 4 add 6M NH_3 until the solution is basic. Then add 0.5 ml more; stir. If copper ion is present, the solution turns blue. A white precipitate in the solution implies that bismuth is present. Centrifuge and decant the solution into a test tube. The solution may contain $Cu(NH_3)_4^{2+}$ and $Cd(NH_3)_4^{2+}$. Wash the precipitate with 1 ml of water and 0.5 ml of 6M NH_3. Stir, and centrifuge and discard the wash.

Step 6. To the precipitate from Step 5 add 0.5 ml of 6M HCl and stir. Any $Bi(OH)_3$ present dissolves to yield Bi^{3+}. A white insoluble residue may contain lead. Centrifuge and decant the solution into a test tube. Wash the precipitate with 1 ml of water and 1 ml of 6M HCl. Centrifuge and discard the wash.

Step 7. *Confirmation of the presence of bismuth.* Add 2 or 3 drops of the decantate from Step 6 to 300 ml of water. A white cloudiness due to precipitation of BiOCl appears if the sample contains bismuth. To the rest of the decantate add 6M NaOH until it is distinctly basic; a white precipitate is $Bi(OH)_3$. To the precipitate add 2 drops of 0.1 M $SnCl_2$ and stir; if bismuth is present, it is reduced to black metallic bismuth.

Step 8. *Confirmation of the presence of lead.* To the white precipitate from Step 6 add 1 ml of 6M acetic acid. Heat gently to dissolve any lead-containing salts. Add 1 ml of water and 1 ml of 6M H_2SO_4. A white precipitate of $PbSO_4$ confirms the presence of lead.

Step 9. To half of the solution from Step 5 add 1 ml of 6M NaOH. The formation of a white precipitate indicates the presence of cadmium. Centrifuge and decant the solution into a test tube. Wash the precipitate twice with 1 ml of water and 1 ml of 6M NH_3, centrifuging each time and discarding the wash.

Step 10. *Confirmation of the presence of copper.* If the solution from Steps 5 and 9 is blue, copper must be present, because the color is characteristic of $Cu(NH_3)_4^{2+}$. To further confirm the presence of Cu(II) add 6M acetic acid to the blue solution from Step 9 until the color fades and the solution becomes acidic. Then add a drop or two of 0.1M $K_4Fe(CN)_6$, which forms a red-brown precipitate of $Cu_2Fe(CN)_6$ if copper is present.

Step 11. *Confirmation of the presence of cadmium.* Dissolve the precipitate from Step 9 in a little 6M HCl; add 6M NaOH until the solution is just basic (a precipitate of $Cd(OH)_2$ should form), and then add 1 ml of 1M thioacetamide. Heat five minutes in the water bath. A yellow precipitate of CdS confirms the presence of cadmium. If the other half of the decantate from Step 5 is blue, add ~0.3 g of solid KCN, which decolorizes the solution, forming complex cyanide ions of cadmium and copper. Then, regardless of whether the solution was blue, add 1 ml of 1M thioacetamide to it and heat it in the water bath; the precipitation of yellow CdS confirms the presence of cadmium.

Step 12. To the decantate from Step 3 add 6M HCl, drop by drop, until the solution is acidic. The sulfides of Sb(V), Sn(IV), and Hg(II) reprecipitate if these cations are present; the precipitate will probably be brown. Add 0.5 ml of 1M thioacetamide and heat on the boiling water bath for three minutes. Centrifuge and discard the liquid. Wash the precipitate with 2 ml of water, and centrifuge and discard the wash.

Step 13. To the precipitate from Step 12, which may contain Sb_2S_5, SnS_2, and HgS, add 1 ml of 12M HCl. Put the test tube in the boiling water bath; heat for five minutes, with stirring. Add 1 ml of 1M thioacetamide and heat for one more minute. Centrifuge out any black residue, which is probably HgS. Decant the liquid, which may contain antimony and tin, into a 30 ml beaker. Wash the precipitate with 1 ml of water and 1 ml of 6M HCl; centrifuge and discard the wash.

Step 14. To the precipitate from Step 13 add 1 ml of 6M HCl and 1 ml of 6M HNO_3. Put the test tube in the water bath for two minutes. The black residue should dissolve, probably leaving some yellow insoluble sulfur. Centrifuge and pour the liquid into a 30 ml beaker, discarding any solid residue. Add 1 ml of water to the liquid in the beaker and boil gently for about a minute.

Step 15. *Confirmation of the presence of mercury.* Put a drop of the liquid from Step 14 on a copper penny or immerse a piece of copper wire in the liquid for a few seconds; a shiny deposit of liquid mercury confirms the presence of mercury. Pour half of the remaining liquid into a test tube and add 1 or 2 drops of 0.1M $SnCl_2$; a white, somewhat glossy, precipitate of Hg_2Cl_2 again shows the presence of mercury. To the rest of the liquid add 2 or 3 drops of 0.1M KI. If no red precipitate of HgI_2 forms, add 3 more drops. Formation of the red precipitate confirms the existence of mercury in the solution.

Step 16. The solution of tin and antimony from Step 13 may contain a trace of bismuth, and possibly other cations from Group II, which have dark sulfides. Boil the solution gently for two minutes to remove H_2S. Add 1 ml of water, transfer the solution to a test tube, and then add 6M NaOH, drop by drop, until the solution is distinctly basic. If a precipitate forms in the solution while it is still acidic and then dissolves in excess base, antimony is present. Any white residue is mostly $Bi(OH)_3$. Centrifuge and pour the liquid into a test tube, discarding the residue. Add 6M HCl, drop by drop, until the solution is acidic to litmus; then add 1 ml more, regenerating the chloro complex ions of tin and antimony in the solution.

Step 17. *Confirmation of the presence of tin.* To half of the solution from Step 16 add a 2 cm length of 24 gauge aluminum wire. This reacts with the acid, producing H_2 and reducing any tin present to Sn^{2+} and any antimony to the metal, which appears as black specks. Centrifuge out any solid after the aluminum wire is all reacted; decant the liquid into a test tube. To the liquid add a drop or two of 0.1M $HgCl_2$. A white opalescent precipitate of Hg_2Cl_2, which may turn gray with time, establishes the presence of tin.

Step 18. *Confirmation of the presence of antimony.* To the other half of the solution from Step 16 add NaOH to bring the pH to 0.5. If tin is present, add about 0.5 g of oxalic acid, and stir until no more crystals dissolve. Then add 1 ml of 1M thioacetamide and put the test tube in the boiling water bath. Precipitation of red-orange Sb_2S_3 confirms the presence of antimony.

COMMENTS ON PROCEDURE FOR ANALYSIS
OF GROUP II CATIONS

Step 1. Here the solution is concentrated, and all ions are brought to their higher state of oxidation. Tin(II) and antimony(III), if present, are oxidized to tin(IV) and antimony(V). This is important, particularly in the case of tin, because unless SnS_2 is the sulfide precipitated, tin will not behave as it should in a later step in the procedure.

Step 2. At this point all the Group II sulfides are precipitated with hydrogen sulfide. The reaction that occurs is in all cases similar to that for copper:

$$Cu^{2+} + H_2S \longrightarrow CuS(s) + 2\,H^+$$

Because the solubility products of the Group II sulfides are very low, they precipitate even when the S^{2-} concentration is very low. The precipitation is carried out at a pH of 0.5, where the $[S^{2-}]$ is only about 1×10^{-20}M. The actual values of the solubility products of some common metallic sulfides are as follows:

GROUP II		GROUP III	
HgS	1×10^{-52}	CoS	1×10^{-21}
CuS	6×10^{-36}	NiS	1×10^{-22}
CdS	1×10^{-26}	ZnS	1×10^{-23}
PbS	1×10^{-27}	FeS	6×10^{-18}

Under the conditions of the precipitation the metallic cations are all about 0.01M; therefore $[M^{2+}][S^{2-}] \approx (1 \times 10^{-2})(1 \times 10^{-20}) \approx 1 \times 10^{-22}$. This means that those in the left column, $K_{sp} < 10^{-22}$, precipitate, whereas those in the right column, $K_{sp} > 10^{-22}$, do not. This difference in sulfide solubilities is the basis of the separation of Group II cations (left column) from Group III cations (right column).

Clearly, if the separation is to be effective, the $[H^+]$ must be properly set before precipitation is carried out. This is perhaps most easily done by using an indicator to establish the pH of the solution. We find that methyl violet is quite suitable. On a piece of filter paper put down about 10 drops of methyl violet indicator, making 10 spots. Let these dry in the air. Make up a solution of HCl about 0.3M in H^+ ion by diluting 10 ml of a 1M HCl solution with 20 ml of water and stirring. Put a drop of this acid on a drop of indicator. The green-blue color indicates the desired pH of 0.5. Add 6M HCl, drop by drop, to your cation solution, with stirring, until the color of the indicator when touched with a drop of solution from your stirring rod matches that of the standard spot.

The source of H_2S in our procedure is thioacetamide, CH_3CSNH_2. This organic compound in either acidic or basic solution tends to hydrolyze when heated, producing H_2S; in acid the reaction is:

$$CH_3CSNH_2 + 2\,H_2O + H^+ \longrightarrow CH_3COOH + NH_4^+ + H_2S$$

Generating H_2S in small amounts by this reaction is advantageous, because the gas is both bad-smelling and highly toxic. Also, it is produced slowly, which tends to allow more compact sulfide precipitates to form. One difficulty that is sometimes encountered with this precipitation is that CdS, which is somewhat more soluble than are the other Group II sulfides, does not always precipitate properly.

The pH of the solution, at least for some cation mixtures, tends to decrease significantly during the precipitation, making it necessary to test the pH of the liquid after decantation from the sulfides. In some trials, readjusting the pH if necessary, adding more thioacetamide, and heating cause precipitation of the yellow CdS. It is essential that you remove Cd^{2+} from the solution at this stage, because if you do not precipitate it, you will miss detecting cadmium and you will also contaminate the solution of Group III and Group IV cations if they are present.

Step 3. This procedure separates the sulfides of copper, bismuth, cadmium, and lead from those of tin, antimony, and mercury. The former do not tend to form sulfide complexes, whereas the latter readily do under the alkaline conditions that prevail; for SnS_2 the reaction is:

$$SnS_2(s) + S^{2-} \rightleftharpoons SnS_3^{2-}$$

The compound SnS will not undergo this reaction; this is the reason it is important to oxidize Sn(II) to Sn(IV) in Step 1. Under the conditions of the separation, some of the tin may also be present as a hydroxide complex ion, but this causes no difficulty. Most of the HgS dissolves during this step, and, unfortunately, a trace of the Bi_2S_3 also goes into solution.

Step 4. The sulfides of copper, bismuth, cadmium and tin dissolve readily in hot 6M HNO_3, a typical reaction being:

$$Bi_2S_3(s) + 2\,NO_3^- + 8\,H^+ \longrightarrow 2\,Bi^{3+} + 3\,S(s) + 2\,NO(g) + 4\,H_2O$$

Any HgS present is not affected by the nitric acid.

Step 5. The blue copper ammonia complex ion forms at this point, with the color being definitive for the presence of copper. A white precipitate simultaneously formed is highly indicative of bismuth, although if lead is present it also precipitates. Cadmium and copper form stable ammonia complex ions, whereas bismuth and lead do not; typically:

$$Cd^{2+} + 4\,NH_3 \rightleftharpoons Cd(NH_3)_4{}^{2+}$$
$$Bi^{3+} + 3\,NH_3 + 3\,H_2O \rightleftharpoons Bi(OH)_3(s) + 3\,NH_4{}^+$$

Step 6. Lead, if present in Step 5, precipitates as a highly insoluble basic salt. This salt is essentially insoluble in 6M HCl, possibly being converted to the insoluble chloride, and therefore can be readily separated from the bismuth hydroxide:

$$Bi(OH)_3(s) + 3\,H^+ \rightleftharpoons Bi^{3+} + 3\,H_2O$$

Step 7. There are two very good confirmatory tests for bismuth. Simply pouring a drop or two of the acidic solution of Bi^{3+} into water causes a characteristic white cloudiness because of BiOCl. Alternatively, if you make the bismuth solution basic with NaOH and add a drop of $SnCl_2$ solution, an oxidation-reduction reaction occurs, forming black metallic bismuth. You may use either, or both, of these tests to confirm the presence of bismuth.

Step 8. In all probability, you do not have appreciable amounts of lead precipitating in Group II analysis, because the chloride precipitation in Group I analysis is really quite effective. However, if you do have an insoluble white residue in Step 6, it is probably a lead salt. You may confirm this by heating the salt with some acetic acid in which the salt is soluble, and then precipitating lead sulfate by addition of sulfuric acid. Bismuth does not interfere with this test.

Steps 9, 10, 11. Detection of cadmium in the presence of copper is accomplished in many procedures by complexing the copper and cadmium with cyanide ion and then precipitating cadmium sulfide, which can be precipitated from the solution, whereas the copper cyanide complex ion is too stable for CuS to form under such conditions. There are two drawbacks to this approach. One is that cyanides are highly poisonous and perhaps should be avoided with beginning students, and the other is that the precipitate obtained tends to be dark and may mask any yellow CdS that may be present.

We have found that, if one adds 6M NaOH to the solution of ammonia complex ions, a precipitate forms if cadmium is present. This precipitate is insoluble in ammonia, and therefore can be washed free of the copper complex ion. The formation of such a precipitate is highly indicative of the presence of cadmium. The precipitate is dissolved in HCl, the pH is adjusted, and then yellow CdS is precipitated on addition of thioacetamide and application of heat. This approach avoids the use of cyanide, and in our opinion also gives a better confirmatory test for cadmium.

If a blue solution is obtained in Step 5 or Step 9, it can be taken as a confirmatory test for copper. A simple additional test is to acidify a portion of the solution from Step 9 with acetic acid and to add a drop of potassium ferrocyanide solution; if the solution contains copper, a red-brown precipitate of $Cu_2Fe(CN)_6$ forms.

Step 12. The sulfide complexes of tin, antimony, and mercury are destroyed by acidification and the sulfides are reprecipitated; typically, the following reaction occurs:

$$HgS_2^{2-} + 2\,H^+ \rightleftharpoons HgS(s) + H_2S$$

Step 13. The sulfides of tin and antimony are sufficiently soluble in acid, and the chloro complex ions of these elements are sufficiently stable, that the sulfides dissolve readily in 12M HCl at 100° C. Mercuric sulfide is not soluble in this solution and is recovered as a finely divided black solid; for antimony(V) sulfide the reaction is:

$$Sb_2S_5(s) + 12\,Cl^- + 6\,H^+ \longrightarrow 2\,SbCl_6^{3-} + 3\,H_2S + 2\,S(s)$$

Antimony(V) is a sufficiently strong oxidizing agent to produce some free sulfur in this reaction; note that the antimony is reduced to the +3 state.

Step 14. Mercuric sulfide is readily soluble in aqua regia. This solvent here is a rather mild aqua regia, but it brings the HgS into solution easily without actual boiling:

$$3\,HgS(s) + 12\,Cl^- + 2\,NO_3^- + 8\,H^+ \longrightarrow$$
$$3\,HgCl_4^{2-} + 2\,NO(g) + 4\,H_2O + 3\,S(s)$$

The residue from this reaction is a small amount of yellow sulfur, which usually coagulates and floats on the liquid.

Step 15. There are three good confirmatory tests for the mercuric ion. The simplest is to put a drop of the solution from Step 14 on a copper penny

Outline of Procedure for Analysis of Group II Cations

Ions possibly present:

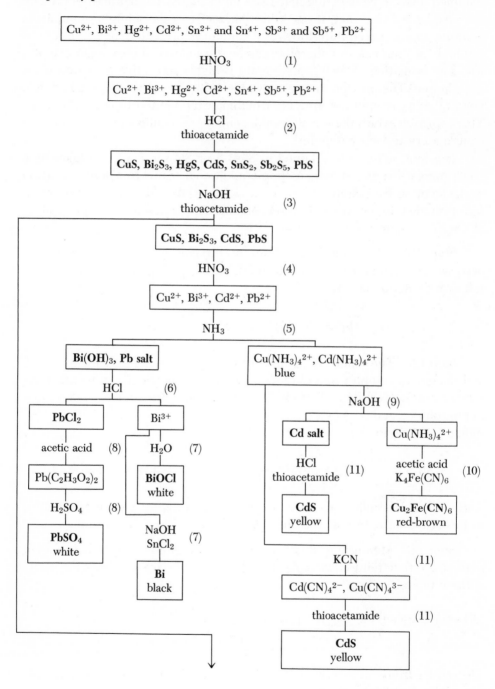

Cu^{2+}, Bi^{3+}, Hg^{2+}, Cd^{2+}, Sn^{2+} and Sn^{4+}, Sb^{3+} and Sb^{5+}, Pb^{2+}

HNO_3 (1)

Cu^{2+}, Bi^{3+}, Hg^{2+}, Cd^{2+}, Sn^{4+}, Sb^{5+}, Pb^{2+}

HCl
thioacetamide   (2)

CuS, Bi_2S_3, HgS, CdS, SnS_2, Sb_2S_5, PbS

NaOH
thioacetamide   (3)

CuS, Bi_2S_3, CdS, PbS

HNO_3 (4)

Cu^{2+}, Bi^{3+}, Cd^{2+}, Pb^{2+}

NH_3 (5)

$Bi(OH)_3$, Pb salt

$Cu(NH_3)_4{}^{2+}$, $Cd(NH_3)_4{}^{2+}$
blue

HCl   (6)

$PbCl_2$

Bi^{3+}

acetic acid   (8)

H_2O (7)

$Pb(C_2H_3O_2)_2$

$BiOCl$
white

H_2SO_4 (8)

$PbSO_4$
white

NaOH
$SnCl_2$ (7)

Bi
black

NaOH   (9)

Cd salt

$Cu(NH_3)_4{}^{2+}$

HCl
thioacetamide   (11)

acetic acid
$K_4Fe(CN)_6$ (10)

CdS
yellow

$Cu_2Fe(CN)_6$
red-brown

KCN   (11)

$Cd(CN)_4{}^{2-}$, $Cu(CN)_4{}^{3-}$

thioacetamide   (11)

CdS
yellow

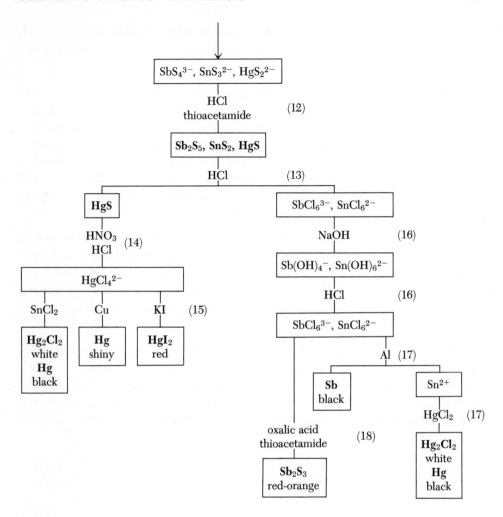

or to dip a piece of heavy copper wire in the solution. If mercury is present, it quickly deposits as a bright shiny metal on the copper surface:

$$HgCl_2 + Cu(s) \longrightarrow Hg(l) + Cu^{2+} + 2\ Cl^-$$

Probably the most common test is to add a drop or two of $SnCl_2$ solution to the solution under scrutiny; if Hg(II) is present, it is reduced to the mercurous state or to the black metal:

$$2\ HgCl_2 + Sn^{2+} \longrightarrow Hg_2Cl_2(s) + Sn^{4+} + 2\ Cl^-$$

Finally, if a few drops of dilute KI solution are added to a solution containing mercury(II), a bright red-orange precipitate forms, which is soluble in excess iodide:

$$HgCl_2 + 2\ I^- \longrightarrow HgI_2(s) + 2\ Cl^-$$
$$HgI_2(s) + 2\ I^- \rightleftharpoons HgI_4^{2-}$$

The Hg(II) ion in chloride solution exists as undissociated $HgCl_2$, which explains what may appear to be anomalous formulas in the preceding ionic equations.

Step 16.   In the separation of tin, antimony, and mercury from the other Group II cations, a certain amount of the other species, mainly bismuth(III), is also dissolved. This must be removed if one is to get a light-colored product in the sulfide test for antimony in a later step. Concentrated NaOH precipitates Bi(III) but not the tin and antimony, because their hydroxides are amphoteric. If antimony is present, a precipitate forms even while the solution is acidic. This dissolves in the excess base; formation of a precipitate behaving this way is a strong indication for antimony.

Step 17.   In acid solution aluminum reduces antimony(III) to the metal. Bismuth is also reduced to a black solid under these conditions; therefore it is highly advisable that it not be present. Tin is also reduced to the metal if the concentration of acid is low, but under the conditions here, it goes to the Sn(II) state. The confirmatory test is again the reaction of $HgCl_2$ with Sn(II).

Step 18.   Indications of the presence of antimony appear several times in the procedure. If tin is present, it may be complexed by oxalic acid and Sb_2S_3 then precipitated. If tin is not present, you may simply adjust the pH of the solution from Step 16 and add thioacetamide. The precipitate is the correct color if you have removed the trace impurities properly.

PROBLEMS

1. Write balanced net ionic equations to explain the following observations:
   a. A precipitate forms when H_2S is bubbled into a solution containing Bi^{3+}.
   b. Addition of NH_3 to a solution of Cu^{2+} gives a light blue precipitate, which then dissolves to give a deep blue solution (two reactions).
   c. Cadmium sulfide dissolves in 6M H_2SO_4.
   d. A precipitate forms when NaOH is added to a solution containing Hg^{2+}.
   e. Antimony(III) sulfide dissolves in hot concentrated HCl.

2. Write balanced net ionic equations for the following oxidation-reduction reactions:
   a. Copper(II) sulfide dissolves in concentrated HNO_3.
   b. Bismuth hydroxide is reduced to the metal by Sn^{2+} in basic solution.
   c. Mercury(II) sulfide dissolves in aqua regia, a mixture of 6M HCl and 6M HNO_3.
   d. Aluminum metal reduces Sn^{4+} to Sn^{2+}.

3. Which of the hydroxides $Cu(OH)_2$, $Bi(OH)_3$, $Cd(OH)_2$, and $Pb(OH)_2$ dissolve in an excess of the following:
   a. NH_3?   b. NaOH?

4. Suggest a reagent that brings into solution:
   a. Ag_2S b. CuS   c. SnS   d. CdS

5. Describe simple tests that would enable you to distinguish between the following:
   a. $AgNO_3(s)$ and $Cu(NO_3)_2(s)$ e. Cu^{2+} and Sn^{2+}
   b. $Cd(OH)_2(s)$ and $Bi(OH)_3(s)$ f. Bi^{3+} and Cd^{2+}
   c. $CuS(s)$ and $SnS_2(s)$ g. Hg^{2+} and Cu^{2+}
   d. $SnCl_2(s)$ and $BiCl_3(s)$ h. Ag^+ and Sn^{4+}

6. Indicate how you would accomplish the following conversions.
   a. $Bi(NO_3)_3(s) \longrightarrow Bi_2S_3(s)$ d. $Na_2S(s) \longrightarrow NaCl(s)$
   b. $CuSO_4(s) \longrightarrow CuI(s)$ e. $Hg^{2+} \longrightarrow Hg(l)$
   c. $Cd(NO_3)_2(s) \longrightarrow Cd(OH)_2(s)$

7. State precisely what would be observed in each step of the Group II analysis with an unknown containing only the following cations:
   a. Cu^{2+}, Hg^{2+}, and Sn^{2+} b. Bi^{3+}, Cd^{2+}, and Sb^{3+}

8. Explain why:
   a. The solution in Step 1 is boiled with HNO_3.
   b. The Pb^{2+} cation is frequently carried over from Group I to Group II.
   c. Nitric acid is added in Step 4.
   d. Sulfur is frequently left as a residue when a metal sulfide is treated with HNO_3.
   e. The formation of a black precipitate in Step 17 is not conclusive evidence for Sb^{3+}.

9. What effect would the following "errors" in procedure have on the results of an analysis of a Group II unknown?
   a. In Step 1, the solution was too basic.
   b. In Step 5, NaOH was added instead of NH_3.
   c. In Step 7, the decantate was added to dilute HCl instead of to water.
   d. Step 16 was omitted.

10. Suppose you were told that a Group II unknown could contain only the following cations. Work out an abbreviated scheme of analysis for these ions, omitting extra steps.
    a. Cu^{2+}, Bi^{3+}, Cd^{2+} b. Cu^{2+}, Hg^{2+}, Sb^{3+}

11. Describe schemes of analysis, as concise as possible, for the following ions:
    a. Ag^+, Bi^{3+}, Sb^{3+} b. Pb^{2+}, Cu^{2+}, Hg^{2+}

12. On the basis of the following observations, state whether each of the ions in Group II is present, absent, or questionable.
    The precipitate obtained with H_2S in acidic solution is black. This precipitate is completely soluble in 6M NaOH.

13. From the following observations, state whether each of the Group II ions is present, absent, or questionable.
    The sulfide precipitate is partially soluble in NaOH. The residue from the NaOH treatment is dissolved in HNO_3; addition of NH_3 gives a colored solution and a white precipitate. The NaOH solution is acidified to give a precipitate that is completely soluble in 12M HCl. The HCl solution is treated as in Step 16; a piece of aluminum wire is then added. The resultant solution gives a precipitate upon addition of $HgCl_2$.

14. A solid unknown may contain any of the following:

$$AgCl, \ Cu(OH)_2, \ Bi_2S_3, \ Pb(NO_3)_2$$

The solid is insoluble in water but completely soluble in ammonia to give a deeply colored solution. Identify, as completely as possible, the solids that are present. Suggest additional tests that one might use to further identify the unknown.

101. Write balanced net ionic equations for the reactions that occur when:
     a. Solutions of $CuSO_4$ and $K_4Fe(CN)_6$ are mixed.
     b. An acidic solution of $BiCl_3$ is diluted with water.
     c. Aqueous ammonia is added to a solution containing Cd^{2+} (two reactions).
     d. Iodide ions are added to a solution containing Hg^{2+} (two reactions).
     e. Tin(IV) sulfide is treated with concentrated NaOH.

102. Write balanced net ionic equations for the following oxidation-reduction reactions:
     a. Solutions of $CuSO_4$ and KI are mixed.
     b. Cadmium sulfide is treated with HNO_3. (The reaction is similar to that with CuS.)
     c. Solutions of $Hg(NO_3)_2$ and $SnCl_2$ are mixed; a white precipitate forms, and Sn(IV) is produced.
     d. Tin is treated with concentrated nitric acid.

103. Which of the ions Sn^{2+}, Sn^{4+}, and Sb^{3+} form hydroxides that dissolve in an excess of the following:
     a. NH_3?   b. NaOH?
     Write net ionic equations for any reactions that occur.

104. Write balanced net ionic equations for reactions by which sulfides of the following ions can be brought into solution:
    a. Pb^{2+} b. Bi^{3+} c. Hg^{2+} d. Sb^{3+}

105. Describe simple tests to distinguish between the following:
    a. $CuS(s)$ and $Bi_2S_3(s)$ e. Sn^{2+} and Sn^{4+}
    b. $Cd(OH)_2(s)$ and $CdS(s)$ f. Pb^{2+} and Bi^{3+}
    c. $AgCl(s)$ and $SnO_2(s)$ g. S^{2-} and SO_4^{2-}
    d. $HgI_2(s)$ and $NaI(s)$ h. Bi^{3+} and Sb^{3+}

106. Indicate how you would accomplish the following conversions:
    a. $CuS(s) \longrightarrow Cu(NO_3)_2(s)$ d. $Sn^{2+} \longrightarrow Sn^{4+}$
    b. $Bi^{3+} \longrightarrow Bi(s)$ e. $Sn^{4+} \longrightarrow Sn^{2+}$
    c. $NaOH(s) \longrightarrow NaNO_3(s)$

107. State precisely what would be observed in each step of the Group II analysis with an unknown containing only the following cations:
    a. Hg^{2+}, Sn^{4+}, and Sb^{3+} b. Pb^{2+}, Cu^{2+}, and Sn^{2+}

108. Explain why:
    a. The addition of H_2S in *acidic* solution separates Group II from Group III cations.
    b. Unlike CuS, SnS_2 dissolves in strongly basic solution.
    c. The formation of a white precipitate in Step 5 is not conclusive evidence for Bi^{3+}.
    d. Acetic acid is used in Step 8.
    e. Oxalic acid is used in Step 18.

109. What effect, if any, would the following "errors" in procedure have on the analysis of a Group II unknown?
    a. In Step 1, the solution was too acidic.
    b. Step 3 was omitted.
    c. In Step 9, NH_3 was used instead of NaOH.
    d. In Step 14, HNO_3 was omitted.

110. An unknown can contain only the following cations. Work out an abbreviated scheme of analysis for these ions, omitting any unnecessary steps.
    a. Sn^{2+}, Hg^{2+}, Sb^{3+} b. Bi^{3+}, Sn^{4+}, Sb^{5+}

111. Work out abbreviated schemes of analysis for unknowns that can contain no cations other than the following:
    a. Hg_2^{2+}, Cd^{2+}, Sn^{2+} b. Ag^+, Cu^{2+}, Sn^{2+}, Sb^{3+}

112. Based on the following observations, state whether each of the ions in Group II is present, absent, or questionable.
    The sulfide precipitate is light in color; there is no evidence for the formation of a black sulfide. The precipitate is partially soluble in NaOH.

113. From the following observations, state whether each of the ions in Group II is present, absent, or questionable.
    The sulfide precipitate is partially soluble in NaOH. The residue resulting from treatment with NaOH dissolves in nitric acid; addition of ammonia to the solution leaves a colorless solution and no precipitate. Acidification of the NaOH solution gives a precipitate that is completely insoluble in 12M HCl.

LABORATORY ASSIGNMENTS

Perform one or more of the following, as directed by your instructor:

1. Using a known sample containing about 1 ml of 0.1M nitrate solutions of each of the Group II cations, analyze the solution by the procedure given in this text. Compare your observations with those described. Obtain a Group II unknown and analyze it by the procedure. Report your observations and conclusions on a Group II flow chart.

2. You will be given an unknown containing one cation from Group I and one cation from Group II. Develop as simple a scheme as you can for analyzing this limited unknown. Draw a flow diagram, including all the reagents used and the observations to be expected in case any given cation is present. Show your procedure to your instructor for approval. Obtain an unknown and analyze it by your procedure, using another color on the flow chart to note your observations and conclusions at each step. When you believe you know which two cations are present, make up a known solution containing those cations and test it to see that it behaves in the same way as your unknown.

3. You will be given an unknown that may contain only two cations, both chosen from Group II. You will be told which two cations may be present. Develop a scheme by which a solution containing those ions might be analyzed. When you are sure that your scheme will work, use it to analyze an unknown you obtain from your instructor. Report your results for evaluation, noting the observations that lead you to believe your analysis to be correct. Your instructor will then assign you another pair of Group II cations to consider. Your grade will depend on correct analyses of the unknowns you are assigned and the number of different analyses you are able to complete in the time allowed. Some possible pairs of cations you may be assigned are as follows:

Cu^{2+}, Hg^{2+}	Cd^{2+}, Sb^{3+}	Sn^{4+}, Hg^{2+}
Cd^{2+}, Sn^{4+}	Pb^{2+}, Cu^{2+}	Bi^{3+}, Hg^{2+}
Bi^{3+}, Pb^{2+}	Sn^{4+}, Sb^{3+}	Bi^{3+}, Cd^{2+}

4. Several days in advance, your instructor will assign you an unknown that may contain only the cations indicated in one of the following sets. Before coming to the laboratory, develop a procedure by which you can analyze the unknown for the ions in your set. Do not include any unnecessary steps, and try to make your scheme for separation as simple as possible. Draw a flow chart for your procedure, indicating all reagents to be used and all species that may be formed. You may find that your procedure is not at all like that given for a general Group II unknown. Test your scheme with a known sample containing 1 ml of 0.1M nitrate solutions of each of the cations in your set. When you are sure your procedure works, use it to determine the cations present in an unknown furnished by your instructor. Your set of cations will be one of the following:

a. Cu^{2+}, Cd^{2+}, Pb^{2+}, Sn^{2+} d. Hg^{2+}, Pb^{2+}, Cd^{2+}, Sn^{4+}
b. Bi^{3+}, Hg^{2+}, Sb^{3+}, Cd^{2+} e. Bi^{3+}, Pb^{2+}, Hg^{2+}, Sb^{5+}
c. Cu^{2+}, Hg^{2+}, Sn^{4+}, Sb^{3+} f. Cu^{2+}, Bi^{3+}, Cd^{2+}, Sn^{4+}

SECTION 3

THE PROPERTIES OF THE CATIONS IN GROUP III— Fe^{2+} and Fe^{3+}, Al^{3+}, Zn^{2+}, Mn^{2+}, Cr^{3+}, Co^{2+}, Ni^{2+}

IRON, Fe^{2+} AND Fe^{3+}

Iron in its compounds is ordinarily either in the $+2$, ferrous, or the $+3$, ferric, state. The latter state is more common, because most ferrous compounds oxidize in air, particularly in the presence of water. Iron(II) salts are typically found as hydrates and are light green. The two most common examples are sulfates, $FeSO_4 \cdot 7\,H_2O$ and $Fe(NH_4)_2(SO_4)_2 \cdot 6\,H_2O$. Iron(III) salts are also ordinarily hydrates and have various colors. Although in nitric acid solution the hydrated iron(III) ion, $Fe(H_2O)_6^{3+}$, is essentially colorless, hydrolysis in less acid systems results in a yellow or amber-colored Fe(III) species. The hydrate $FeCl_3 \cdot 6\,H_2O$ is yellow; $Fe_2(NH_4)_2(SO_4)_4 \cdot 24\,H_2O$ is light violet; anhydrous ferric sulfate, $Fe_2(SO_4)_3$, is white.

In either oxidation state iron forms many complex ions, perhaps the most stable being the cyanide complexes of iron(II) and iron(III), $Fe(CN)_6^{4-}$ and $Fe(CN)_6^{3-}$. These ions are commonly called ferrocyanide and ferricyanide; their official names are hexacyanoferrate(II) and hexacyanoferrate(III), respectively.

CHARACTERISTIC REACTIONS OF THE IRON(II) AND IRON(III) IONS

1. **Soluble Sulfides.** In slightly alkaline solution, sulfide ion precipitates black FeS from solutions of Fe(II) and of Fe(III):

$$Fe^{2+} + S^{2-} \rightleftharpoons FeS(s)$$
$$2\,Fe^{3+} + 3\,S^{2-} \longrightarrow 2\,FeS(s) + S(s)$$

The sulfide of iron is readily soluble in 6M HCl, which is completely consistent with the fact that the sulfide cannot be precipitated in acid solution:

$$FeS(s) + 2\,H^+ \rightleftharpoons Fe^{2+} + H_2S$$

2. **Sodium Hydroxide and Ammonia.** These reagents precipitate hydroxides; iron(II) forms white, gelatinous, easily oxidized ferrous hydroxide, $Fe(OH)_2$, whereas red-brown gelatinous ferric hydroxide, $Fe(OH)_3$, is the product from Fe(III) salts. Ferric hydroxide is much less soluble than is the ferrous compound, but both dissolve readily in acid. Neither is soluble in excess of either hydroxide ion or ammonia. $Fe(OH)_3$ tends to coprecipitate other cations, especially those in a $+3$ state, which sometimes complicates its separation from other cations.

3. **Potassium Ferrocyanide.** With Fe^{3+} this reagent produces a highly insoluble, dark blue precipitate:

$$K^+ + Fe^{3+} + Fe(CN)_6{}^{4-} \longrightarrow KFe[Fe(CN)_6](s)$$

The Fe^{2+} ion under these conditions forms a white precipitate, which on exposure to air is slowly converted to the blue compound. This reagent offers an extremely sensitive test for iron(III).

4. **Potassium Thiocyanate.** A solution treated with this reagent turns deep red if Fe^{3+} is present, because of the formation of an Fe(III) thiocyanate complex ion:

$$Fe^{3+} + SCN^- \rightleftharpoons FeSCN^{2+}$$

This reaction is also a highly sensitive test for the presence of Fe^{3+}, and other cations, including Fe^{2+}, do not interfere. The test is usually carried out under slightly acid conditions in which the hydrolysis of Fe^{3+} is minimized. (The actual formula of the complex depends on the relative concentrations of Fe^{3+} and SCN^-. The complex $Fe(SCN)_6{}^{3-}$ is likely to form if $[SCN^-]$ is high.)

5. **Oxidizing and Reducing Agents.** Fe^{2+} ion is, as we have noted, easily oxidized to Fe^{3+} by air; either H_2O_2 in acid or strong nitric acid can be used if rapid oxidation is needed. The reagents H_2S, $SnCl_2$, and KI all reduce Fe^{3+} to Fe^{2+} under acid conditions; typically:

$$2\ Fe^{2+} + H_2O_2 + 2\ H^+ \longrightarrow 2\ Fe^{3+} + 2\ H_2O$$
$$2\ Fe^{3+} + 2\ I^- \longrightarrow 2\ Fe^{2+} + I_2$$

Metallic iron is a good reducing agent. It dissolves readily in dilute HCl and H_2SO_4 with the evolution of hydrogen and the formation of Fe^{2+} ion. It also dissolves in an oxidizing acid, such as dilute HNO_3; in this case the iron is oxidized to Fe^{3+}, and the $NO_3{}^-$ ion rather than the H^+ ion is reduced. Iron is passive in concentrated HNO_3.

ALUMINUM, Al^{3+}

Aluminum in its compounds is ordinarily found only in the $+3$ state. Aluminum salts are colorless and soluble in water and in HCl. These salts are frequently commercially available as hydrates. The aluminum ion in solution, $Al(H_2O)_6^{3+}$, tends to undergo extensive hydrolysis; consequently, solutions in which it is present tend to be acidic. The oxide, Al_2O_3, is refractory; it is soluble in acids and bases unless it has been heated strongly, but under the latter condition it is nearly inert.

CHARACTERISTIC REACTIONS OF THE ALUMINUM ION

1. **Soluble Sulfides.** In slightly basic solution sulfide ion does not precipitate aluminum(III) as a sulfide salt. Instead, we obtain white, gelatinous, very insoluble aluminum hydroxide, $Al(OH)_3$. There is no reaction with H_2S in acid solution. This behavior is completely consistent with the fact that Al_2S_3 prepared by direct reaction of aluminum and sulfur reacts with water to produce $Al(OH)_3$ and H_2S:

$$Al_2S_3(s) + 6\,H_2O \longrightarrow 2\,Al(OH)_3(s) + 3\,H_2S$$

2. **Sodium Hydroxide.** With solutions of strong bases we also obtain a white, gelatinous precipitate of $Al(OH)_3$, which dissolves very easily in excess reagent:

$$Al(OH)_3(s) + OH^- \rightleftharpoons Al(OH)_4^-$$

3. **Ammonia.** The hydroxide $Al(OH)_3$ is more easily obtained from Al(III) solution with this reagent than with NaOH or KOH, because with the latter the soluble complex $Al(OH)_4^-$ is so easily formed. Analytically, aluminum is precipitated with NH_3 in the presence of NH_4^+, which buffers the solution at about the proper pH. When first precipitated, $Al(OH)_3$ is quite soluble in acid as well as basic solution, but after aging it is much less easily dissolved with acid. Detection of $Al(OH)_3$ as a precipitate is facilitated by adding a red dye, aluminon, which tends to be adsorbed from the solution, which is originally pink, on to the precipitate, which becomes red as the solution becomes colorless.

4. **Oxidizing and Reducing Agents.** The Al^{3+} ion is not reduced to the metal in ordinary chemical reactions, because this change requires too much energy. Aluminum metal dissolves easily in dilute strong acids and bases,

with evolution of hydrogen. In oxidizing acids the acid anion, rather than H^+, is usually reduced. In hot nitric acid, surprisingly enough, aluminum is passive.

ZINC, Zn^{2+}

Zinc compounds are typically colorless and are commonly obtained as hydrates. Most zinc compounds are soluble in water and in dilute acids. The zinc ion in solution has the formula $Zn(H_2O)_4{}^{2+}$ and tends to hydrolyze slightly, making its solution acidic:

$$Zn(H_2O)_4{}^{2+} \rightleftharpoons Zn(H_2O)_3(OH)^+ + H^+$$

Zinc(II) forms several complexes, but in other respects its properties are quite similar to those of aluminum. Zinc occurs ordinarily in its compounds only in the $+2$ state.

CHARACTERISTIC REACTIONS OF THE ZINC ION

1. **Soluble Sulfides.** Zinc ion in moderately acidic solution does not react with H_2S. Under alkaline conditions sulfide ion precipitates white ZnS. The sulfide is soluble in dilute strong acids but not in NaOH.

2. **Sodium Hydroxide.** Solutions of strong bases precipitate zinc(II) as a white, gelatinous precipitate of $Zn(OH)_2$, which dissolves very easily in excess reagent to give $Zn(OH)_4{}^{2-}$. The hydroxide is also soluble in strong acids and in ammonia.

3. **Ammonia.** Solutions of NH_3 precipitate white $Zn(OH)_2$, which dissolves by forming a complex ion in excess reagent:

$$Zn(OH)_2(s) + 4\,NH_3 \rightleftharpoons Zn(NH_3)_4{}^{2+} + 2\,OH^-$$

4. **Potassium Ferrocyanide.** Solutions of $K_4[Fe(CN)_6]$ precipitate gray-white $K_2Zn_3[Fe(CN)_6]_2$, which is blue-green if traces of iron are present.

5. **Oxidizing and Reducing Agents.** The zinc ion is not readily reduced, because the metal is a strong reducing agent. Zinc metal reacts with dilute acid solutions, ordinarily yielding gaseous hydrogen. With oxidizing acids, zinc often reduces the anion of the acid in the solution reaction; with dilute nitric acid the $NH_4{}^+$ ion is formed by reduction of $NO_3{}^-$ ion.

MANGANESE, Mn^{2+}

Manganese exists in its common compounds in perhaps more different oxidation states than any other element, i.e., $+2$, $+3$, $+4$, $+6$, and $+7$, but for many purposes, one can restrict his attention to the $+2$, $+4$, and $+7$ oxidation states. The only cation of manganese is Mn^{2+}. In all other oxidation states it is found as a complex anion, e.g., MnO_4^{2-}. The usual Mn(IV) compound encountered is MnO_2, a brown or black oxide that is insoluble in water. Potassium permanganate, $KMnO_4$, a deep purple substance that is soluble in water and that is a strong oxidizing agent in acid solution, contains manganese in its highest oxidation state, $+7$.

Manganese compounds are colored, ranging from pink, characteristic of manganous, Mn(II), salts, to the green of Mn(VI) anions (e.g., MnO_4^{2-}), to the purple of the MnO_4^- ion, and to some very dark oxides. In spite of the color, manganese does not form as many well defined complexes as do many other colored transition metal ions.

CHARACTERISTIC REACTIONS OF THE MANGANOUS, Mn(II), ION

1. **Soluble Sulfides.** These reagents in slightly alkaline solution precipitate Mn(II) as pink MnS. Because this substance is readily soluble in dilute acid, it cannot be prepared by adding H_2S to Mn^{2+} salts under acidic conditions.

2. **Sodium Hydroxide and Ammonia.** These solutions precipitate white, slightly soluble $Mn(OH)_2$, which tends to darken in the air because of oxidation to compounds of Mn(III) and Mn(IV).

3. **Oxidizing and Reducing Agents.** There are several important oxidation-reduction reactions and reagents and we shall consider them separately.

   Hydrogen Peroxide. In basic solution manganese(II) compounds are oxidized and brown MnO_2 is precipitated:

$$Mn(OH)_2(s) + H_2O_2 \longrightarrow MnO_2(s) + 2\,H_2O$$

In acid solution, H_2O_2 acts as a reducing agent and dissolves MnO_2:

$$MnO_2(s) + H_2O_2 + 2\,H^+ \longrightarrow Mn^{2+} + O_2(g) + 2\,H_2O$$

Without reduction, or oxidation, MnO_2 cannot readily be brought into solution.

   Bromine. In basic solution, Br_2 also oxidizes $Mn(OH)_2$ to MnO_2. Oxidation of Mn(II) to MnO_2 in acid medium is possible, but it requires a very strong oxidizing system, such as concentrated HNO_3 saturated with $KClO_3$.

Sodium Bismuthate. In dilute HNO_3, BiO_3^- oxidizes Mn^{2+} without heating to the purple MnO_4^- ion. This is the most important laboratory use of $NaBiO_3$, a Bi(V) compound; the reaction makes possible the easy identification of manganese in mixtures containing many other cations:

$$2\,Mn^{2+} + 5\,BiO_3^- + 14\,H^+ \longrightarrow 2\,MnO_4^- + 5\,Bi^{3+} + 7\,H_2O$$

CHROMIUM, Cr^{3+}

Chromium is ordinarily encountered in the laboratory in a $+3$ or a $+6$ oxidation state. In Cr(III) salts chromium is usually present as a cation, whereas Cr(VI) species are all anions, typically CrO_4^{2-} and $Cr_2O_7^{2-}$. There are some Cr(II) compounds, but they very easily oxidize and are rarely encountered in solution.

As the name implies, chromium compounds are characteristically colored. Chromium(III) complexes are very common; these complex ions are sometimes slow to undergo ligand exchange, and are therefore called inert or nonlabile, to distinguish them from labile complexes, which come into equilibrium with surrounding ligands in solution almost instantaneously. Chromium(III) complexes are usually green or violet and are soluble in water and in acids. Some common salts are $Cr(NO_3)_3 \cdot 9\,H_2O$ (violet) and $K_2Cr_2(SO_4)_4 \cdot 24\,H_2O$ (red-violet, called chrome alum).

Since Cr^{3+} in solution has many properties in common with Al^{3+} and Fe^{3+}, it is ordinarily oxidized to a Cr(VI) anion during analysis. The anion CrO_4^{2-} is yellow, exists in basic solutions, and forms many insoluble salts with other cations. In acidic medium this ion reacts to form the orange dichromate ion, $Cr_2O_7^{2-}$:

$$2\,CrO_4^{2-} + 2\,H^+ \rightleftharpoons Cr_2O_7^{2-} + H_2O$$

CHARACTERISTIC REACTIONS OF THE CHROMIUM(III) ION

1. **Soluble Sulfides.** In weakly basic solutions, $Cr(OH)_3$, a gelatinous, green precipitate, is obtained. Like Al_2S_3, Cr_2S_3 hydrolyzes completely in water and therefore does not form in the reaction. Predictably, H_2S in acid solution does not react with Cr(III) salts.

2. **Sodium Hydroxide and Ammonia.** These reagents precipitate $Cr(OH)_3$, which is soluble in excess hydroxide ion, forming a green solution containing chromite ion, $Cr(OH)_4^-$. In excess NH_3 some dissolving also occurs, yielding a pink or violet complex ion. On boiling a solution of either complex, $Cr(OH)_3$ reprecipitates.

3. **Oxidizing and Reducing Agents.** Chromic ion can be oxidized to the $+6$ state in either acids or bases. In acid the reaction is carried out under extreme conditions, in concentrated HNO_3 loaded with potassium chlorate, itself a very strong oxidizing agent:

$$2 Cr^{3+} + 6 ClO_3^- + H_2O \longrightarrow Cr_2O_7^{2-} + 6 ClO_2(g) + 2 H^+$$

In moderately basic solution the reaction occurs much more easily, with H_2O_2 or another peroxide being used to bring about the oxidation:

$$2 Cr(OH)_4^- + 3 H_2O_2 + 2 OH^- \longrightarrow 2 CrO_4^{2-} + 8 H_2O$$

Reduction of Cr(VI) ions to form chromic compounds is accomplished in acid solution, using H_2S, HI, or concentrated HCl as the reducing agent:

$$Cr_2O_7^{2-} + 8 H^+ + 3 H_2S \longrightarrow 2 Cr^{3+} + 3 S(s) + 7 H_2O$$

The usual precipitates formed in qualitative tests for the presence of chromium are $PbCrO_4$ or $BaCrO_4$, both of which are yellow. The $PbCrO_4$ dissolves in NaOH and strong acids, but not in acetic acid or ammonia. The $BaCrO_4$ goes into solution in strong acids:

$$2 BaCrO_4(s) + 2 H^+ \rightleftharpoons 2 Ba^{2+} + Cr_2O_7^{2-} + H_2O$$

In nitric acid solution, the $Cr_2O_7^{2-}$ ion reacts with H_2O_2 to form a characteristic blue solution containing a peroxide, which perhaps has the formula CrO_5:

$$Cr_2O_7^{2-} + 4 H_2O_2 + 2 H^+ \longrightarrow 2 CrO_5 + 5 H_2O$$

Chromium peroxide is unstable, and the color fades rather quickly. Ether is sometimes added before the H_2O_2, and the CrO_5 is extracted into the ether phase by shaking, because it is more stable there than in the HNO_3. Chromium metal is hard and crystalline. It goes into solution in HCl and H_2SO_4 but not in HNO_3, in which it is passive.

COBALT, Co^{2+}

Cobalt(II) compounds in water solution are characteristically pink, the color of the hydrated cobalt ion, $Co(H_2O)_6^{2+}$. The common cobalt salts are $CoCl_2 \cdot 6 H_2O$, $Co(NO_3)_2 \cdot 6 H_2O$ and $CoSO_4 \cdot 7 H_2O$. Cobalt(II) forms many complex ions, some of which are nonlabile.

Although the $+2$ state is most common for the simple compounds of cobalt, a large number of inert complexes of cobalt(III) are known.

As with iron, there are three oxides, CoO, Co_2O_3, and Co_3O_4. These oxides dissolve in HCl and all yield cobaltous ion, Co^{2+}.

CHARACTERISTIC REACTIONS OF THE COBALT(II) ION

1. **Soluble Sulfides.** In slightly alkaline solutions containing Co(II), sulfides cause a black precipitate of CoS to be formed. Once prepared, CoS does not redissolve readily in HCl even on heating. It goes into solution in hot nitric acid or aqua regia:

$$3\,CoS(s) + 18\,Cl^- + 2\,NO_3^- + 8\,H^+ \longrightarrow$$
$$3\,CoCl_6^{4-} + 3\,S(s) + 2\,NO + 4\,H_2O$$

   In acid solution, H_2S does not precipitate CoS.

2. **Water.** When heated to boiling, a solution of $CoCl_2$ turns blue. On cooling in air or by addition of water, the color becomes a faint pink. This color change is quite dramatic and indicative of the presence of Co^{2+} in solution. It is caused by a change in the nature of the cobalt complex ion with temperature; in the cold, $Co(H_2O)_6^{2+}$ is present, whereas in the boiling solution $CoCl_3(H_2O)_3^-$ or similar anionic chloro complexes are formed.

3. **Sodium Hydroxide.** This reagent first precipitates Co^{2+} as a blue basic salt, which on heating in excess hydroxide is converted to the slightly soluble pink hydroxide, $Co(OH)_2$, with no amphoteric tendencies. In air, $Co(OH)_2$ is gradually oxidized to dark brown $Co(OH)_3$. Hydrogen peroxide in basic solution hastens the oxidation. The $Co(OH)_3$ is not readily dissolved in acid, but, like MnO_2, readily goes into solution as divalent cobalt if a mild reducing agent, such as H_2O_2, is also present.

4. **Ammonia.** This reagent precipitates $Co(OH)_2$ or a basic salt. There is a tendency for complex ion formation in excess NH_3, but in air the most stable ammonia complex ion is $Co(NH_3)_6^{3+}$, a red Co(III) species.

5. **Ammonium Thiocyanate.** This reagent produces a deep blue complex in Co(II) solutions in ethanol, C_2H_5OH, in which the complex is much more stable than in water:

$$Co^{2+} + 4\,SCN^- \rightleftharpoons Co(SCN)_4^{2-}$$

6. **Potassium Nitrite.** If this reagent is added to a moderately concentrated solution of cobalt ions under slightly acidic conditions, precipitation of a yellow Co(III) compound occurs:

$$Co^{2+} + 7\,NO_2^- + 3\,K^+ + 2\,H^+ \longrightarrow K_3Co(NO_2)_6(s) + NO(g) + H_2O$$

7. **Oxidizing and Reducing Agents.** Some of the reactions involving oxidation of Co(II) species have been mentioned. Reduction of cobalt ions to the metal is not common, because the metal is a reasonably good reducing agent. Binary Co(III) salts, such as $CoCl_3$, in acid solution are unstable to reduction to cobalt(II) species, and hence are very strong oxidizing agents. Cobalt is a hard metal that is stable in air. It dissolves easily in HNO_3 and slowly in HCl and H_2SO_4.

NICKEL, Ni^{2+}

Nickel salts, like those of most of the other members of Group III, are typically colored. The hydrates are all green, the color of the $Ni(H_2O)_6{}^{2+}$ complex ion. Nickel forms many complex ions, many of which have characteristic colors. Although nickel is commonly found only in the $+2$ state, it is possible, particularly under basic conditions, to prepare species with nickel in the $+3$ state.

CHARACTERISTIC REACTIONS OF THE NICKEL ION

1. **Soluble Sulfides.** Nickel(II) precipitates as black NiS in alkaline solutions of sulfide ion. When first precipitated, nickel sulfide tends to be colloidal and difficult to remove from solution. Once formed, NiS is only slightly soluble in HCl, even though it cannot be produced by adding H_2S to an acidic solution of Ni^{2+}. Nickel sulfide dissolves in hot nitric acid or aqua regia.

2. **Sodium Hydroxide.** Strongly basic solutions produce a green, slightly soluble, gelatinous precipitate of $Ni(OH)_2$, which does not dissolve in excess reagent. Under alkaline conditions, $Ni(OH)_2$ can be oxidized to $Ni(OH)_3$ by moderately strong oxidizing agents, such as Cl_2 or Br_2.

3. **Ammonia.** The hydroxide precipitate $Ni(OH)_2$, initially formed on addition of NH_3 to nickel(II) solutions, dissolves readily in excess reagent; the blue $Ni(NH_3)_6{}^{2+}$ ion is produced.

4. **Dimethylglyoxime.** This reagent produces a rose-red precipitate when added to a solution of Ni^{2+}. The test is usually carried out under slightly basic conditions and is extremely sensitive to the presence of nickel. Cobalt ion interferes to some extent with the test because it, too, combines with dimethylglyoxime, producing a dark colored complex ion. Dimethylglyoxime has the formula:

$$CH_3-\overset{}{C}=\overset{..}{N}-O-H$$
$$CH_3-\underset{}{C}=\underset{..}{N}-O-H$$

Each nitrogen atom has a pair of unshared electrons that can be used to form a complex with nickel. The complex, which is uncharged and very insoluble, has the formula:

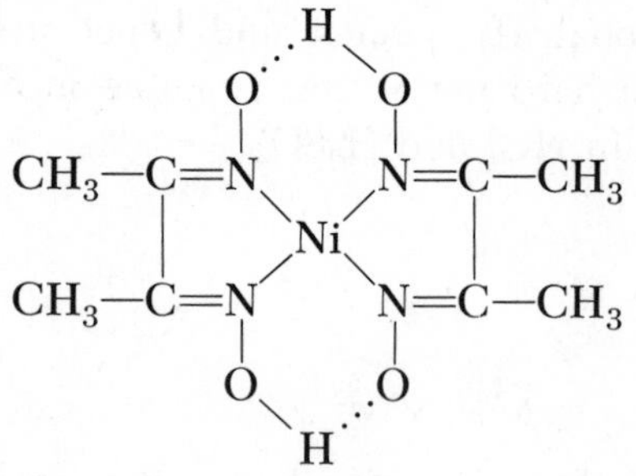

$$
\begin{array}{c}
\text{H} \\
\text{O} \cdots \quad \text{O} \\
\text{CH}_3\text{—C}=\text{N} \qquad \text{N}=\text{C—CH}_3 \\
\qquad \quad \text{Ni} \\
\text{CH}_3\text{—C}=\text{N} \qquad \text{N}=\text{C—CH}_3 \\
\text{O} \quad \cdots \text{O} \\
\text{H}
\end{array}
$$

This is usually abbreviated to $Ni(DMG)_2$.

5. **Oxidizing and Reducing Agents.**   Nickel metal is a moderately good reducing agent; therefore the reduction of Ni^{2+} to the metal does not occur easily. Nickel metal is relatively hard and is stable in air. It is most easily dissolved by dilute HNO_3. Like iron, it becomes unreactive when treated with concentrated nitric acid.

GENERAL DISCUSSION OF THE PROCEDURE FOR ANALYSIS OF THE GROUP III CATIONS Fe^{2+}, and Fe^{3+}, Al^{3+}, Zn^{2+}, Mn^{2+}, Cr^{3+}, Co^{2+}, Ni^{2+}

On looking over the properties of the Group III cations, one finds, as expected, that they do not precipitate in acid solutions containing H_2S. In the discussion of the procedure for precipitating the Group II sulfides, we noted the relatively greater solubility in water, and hence in acid, of the Group III sulfides. However, under slightly alkaline sulfide conditions, the Group III cations precipitate as either sulfides or hydroxides. This property allows one to separate these cations from those in Group IV, none of which precipitates in a sulfide solution at the basic pH used, namely that of an ammonia–ammonium chloride buffer, about pH 9.

Once the cations in Group III have been separated from the solution, the separation of the several ions from one another proceeds rather more easily than does that for Group II cations. The CoS, and for the most part NiS, can be easily removed from the rest of the Group III precipitate because of their relative insolubility in dilute HCl. These two sulfides are then dissolved in aqua regia and identified in each other's presence by reaction with specific reagents. The HCl solution of the other cations, (Fe^{3+}, Al^{3+}, Zn^{2+}, Mn^{2+}, Cr^{3+}) is then oxidized in strongly basic solution, allowing one to separate a precipitate containing $Fe(OH)_3$, MnO_2, and $Ni(OH)_2$ from a solution containing $Al(OH)_4^-$, $Zn(OH)_4^{2-}$, and CrO_4^{2-}. Because MnO_2 is much less basic than are the hydroxides of iron and nickel, it can be separated from them on the basis of its relative insolubility in dilute H_2SO_4. The Fe^{3+} and Ni^{2+} can be readily separated by treatment with ammonia and identified by reagents that give characteristic reactions with those ions.

The solution containing $Al(OH)_4^-$, $Zn(OH)_4^{2-}$, and CrO_4^{2-} is acidified and treated with NH_3, precipitating $Al(OH)_3$ and allowing its removal and identification. Chromate ion is then precipitated as $BaCrO_4$, and zinc ion is detected with a reagent specific to it.

The tests for identification of the ions in Group III have the definite advantage that they tend to be freer of bothersome interferences than are those in Group II. In each case, reagents are used that form very characteristic colored precipitates or complex ions with the given cations. Some of the identifying species and their properties are as follows:

> Fe^{3+}: $FeSCN^{2+}$, a deep red complex ion
> Al^{3+}: $Al(OH)_3$ with aluminon, a red precipitate
> Zn^{2+}: $K_2Zn_3[Fe(CN)_6]_2$, a light green precipitate
> Mn^{2+}: MnO_4^-, a deep purple anion
> Cr^{3+}: CrO_5, a blue solution
> Co^{2+}: $Co(SCN)_4^{2-}$, a blue solution
> Ni^{2+}: Nickel dimethylglyoxime, a rose-red precipitate

PROCEDURE FOR ANALYSIS OF
GROUP III CATIONS

Step 1.   Unless directed otherwise, you may assume that 10 ml of your sample contains the equivalent of about 1 ml of 0.1M solutions of the nitrate or chloride salts of one or more of the Group III cations. If you are working with a general unknown, your sample is the solution decanted after separation of the Group II sulfides. Pour 5 ml of your sample, or your general unknown, into a 30 ml beaker and boil the solution gently until the volume is reduced to about 2 ml.

Pour the solution into a test tube and add 1 ml of 2M NH_4Cl and then, drop by drop, with stirring, add 6M NH_3 until the solution is basic to litmus. Add another 0.5 ml of 6M NH_3. Then add 1 ml of 1M thioacetamide, stir well, and put the test tube in the boiling water bath for at least five minutes, and at least two minutes after the color of the precipitate stops changing. Stir the mixture occasionally as precipitation proceeds.

Centrifuge out the precipitate and decant the liquid into a test tube. Add a few drops of 1M thioacetamide to the liquid and put the test tube into the boiling water bath for a few minutes more to check for complete precipitation of Group III cations. Save the liquid, if necessary, for analysis of Group IV cations. Wash the precipitate twice with 1 ml of 2M NH_4Cl and 2 ml of water, stirring well and centrifuging between washes. Discard the wash liquid.

Step 2.   To the precipitate from Step 1, which should contain the Group III sulfides or hydroxides, add 1 ml of 6M HCl. Mix thoroughly, and pour the slurry into a 30 ml beaker. Boil the liquid gently for about a minute. A black residue is mainly CoS and NiS. Add 1 ml of water, stir, and pour the slurry into a test tube. Centrifuge and decant the liquid, which may contain Al^{3+}, Fe^{2+}, Zn^{2+}, Cr^{3+}, and Mn^{2+}, as well as Ni^{2+}, into a test tube. Wash the solid twice with 1 ml of 6M HCl and 1 ml of water. Centrifuge and discard the wash liquid.

Step 3.   To the precipitate from Step 2 add 1 ml of 6M HCl and 1 ml of 6M HNO_3. Stir it, and put the test tube in the boiling water bath for two minutes. The precipitate should dissolve in a few moments, leaving almost no residue. Pour the solution into a 30 ml beaker and boil gently for one minute; add 1 ml of water and pour the liquid back into a test tube.

Step 4.   *Confirmation of the presence of cobalt.* Pour one third of the solution from Step 3 into a test tube and slowly add about 1 ml of a saturated solution of NH_4SCN in ethyl alcohol, C_2H_5OH. If Co^{2+} is present, a blue solution of $Co(SCN)_4^{2-}$ forms. To the rest of the solution add 6M NaOH, drop by drop, until a precipitate remains after stirring. Then add 0.5 ml of 6M acetic acid to dissolve the precipitate. To half of this solution add 0.4 g solid

KNO_2. Add 1 ml of water and stir. A yellow precipitate of $K_3Co(NO_2)_6$, which may form over a period of ten minutes, again confirms the presence of cobalt.

Step 5. *Confirmation of the presence of nickel.* To the other half of the solution from Step 4, add 2 or 3 drops of dimethylglyoxime reagent. A rose-red precipitate proves the presence of nickel.

Step 6. To the liquid from Step 2, add 1 ml of 3 per cent H_2O_2, and then, drop by drop, with stirring, add 6M NaOH until the solution is basic, and then add 0.5 ml more. Stir the solution for a minute and pour the slurry into a 30 ml beaker. Carefully boil the solution for two minutes. Because this solution tends to bump, scratch the bottom of the beaker continuously with your stirring rod to promote smooth boiling. Transfer the mixture to a test tube and centrifuge out the solid. Decant the liquid, which may contain $Al(OH)_4^-$, $Zn(OH)_4^{2-}$, and CrO_4^{2-}, into a test tube for use in Step 12. Wash the solid once or twice with 1 ml of water and 1 ml of 6M NaOH; centrifuge and discard the wash.

Step 7. To the precipitate from Step 6, which may contain $Fe(OH)_3$, MnO_2, and $Ni(OH)_2$, add 1 ml of water and 1 ml of 6M H_2SO_4. Stir it for two minutes and centrifuge out any undissolved solid, which should contain essentially only MnO_2. Decant the liquid, which may contain Fe^{3+} and Ni^{2+}, into a test tube. Wash the precipitate with 1 ml of water and 1 ml of 6M H_2SO_4; centrifuge and discard the wash.

Step 8. To the precipitate from Step 7 add 1 ml of water and 1 ml of 6M H_2SO_4; stir and then add 1 ml of 3 per cent H_2O_2. Put the test tube in the boiling water bath. The precipitate should dissolve readily on stirring, with possibly a small amount of residue. When it is essentially dissolved, pour the solution into a 30 ml beaker and boil gently for two minutes.

Step 9. *Confirmation of the presence of manganese.* Pour 1 ml of the liquid from Step 8 into a test tube and add 1 ml of 6M HNO_3. Add about 0.2 g of solid sodium bismuthate, $NaBiO_3$, with your spatula, and stir well. There should be a little solid bismuthate in excess. Let the mixture stand for a minute and then centrifuge it. If the solution phase is purple, it is due to MnO_4^- ion and proves the presence of manganese.

Step 10. *Confirmation of the presence of iron.* To half the liquid from Step 7, add 2 ml of water and 1 or 2 drops of 0.1M KSCN. Formation of a deep red solution of the $FeSCN^{2+}$ complex ion is a definitive test for the presence of iron.

Step 11. *Alternative confirmation of the presence of nickel.* To the other half of the liquid from Step 7 add 6M NH_3, drop by drop, until the solution is basic; add 1 or 2 drops more. Centrifuge out any brown precipitate of $Fe(OH)_3$, and decant the liquid into a test tube. Add a few drops of dimethylglyoxime to the liquid. The formation of a rose-red precipitate proves the presence of nickel.

Step 12. Returning to the solution from Step 6, add 6M HNO_3 slowly until the solution is acidic after mixing. Transfer the solution to a 30 ml beaker and boil it down to a volume of about 3 ml. Pour the solution into a test tube. Then add 6M NH_3, drop by drop, until the solution is basic to litmus, and then add 3 drops of NH_3 in excess. Stir the mixture for a minute or so to bring the system to equilibrium. If aluminum is present, a light, translucent, gelatinous precipitate of $Al(OH)_3$ should be floating in the clear (possibly yellow) solution. Centrifuge out the solid, decanting the solution, which may contain CrO_4^{2-} and $Zn(NH_3)_4^{2+}$, into a 30 ml beaker.

Step 13. *Confirmation of the presence of aluminum.* Wash the precipitate from Step 12 with 3 ml of water once or twice, while warming it in the boiling water bath and stirring well. Centrifuge, and discard the wash. Dissolve the precipitate in 0.5 ml of 6M HNO_3, discarding any insoluble residue. Add 1 ml of water and 2 drops of aluminon reagent and stir thoroughly. At this point the *solution* is pink because of the aluminon. Add 6M NH_3, drop by drop, stirring well, until the solution becomes basic to litmus. If Al^{3+} is present, the precipitate of $Al(OH)_3$ re-forms and adsorbs the aluminon from the solution, producing a red *precipitate* of $Al(OH)_3$, a so-called lake, and leaving the previously red solution *essentially colorless*. As you may gather from this discussion, the test for aluminum is *not* the red solution. It is the red *precipitate* of $Al(OH)_3$ and adsorbed aluminon. Centrifuge to concentrate the precipitate and check the color of the solution.

Step 14. If the solution from Step 13 is yellow, chromium is probably present; if it is colorless, chromium is absent. If you suspect that chromium is there, bring the solution to a boil, remove the source of heat, and add 0.5 ml of 1M $BaCl_2$. In the presence of chromium, you obtain a very finely divided yellow precipitate of $BaCrO_4$, which may be mixed with a white precipitate of $BaSO_4$. Pour the slurry into a test tube and put the test tube in the boiling water bath for as much as ten minutes; then centrifuge out the solid and decant the solution into a test tube. Wash the precipitate with 2 ml of water; centrifuge and discard the wash.

Step 15. *Confirmation of the presence of chromium.* To the precipitate from Step 14 add 0.5 ml of 6M HNO_3 and stir to dissolve the $BaCrO_4$. Add 1 ml of water, stir the solution, cool it under the water tap, and then add 2 drops of 3 per cent H_2O_2. A blue solution, which may fade quite rapidly, is confirmatory evidence for the presence of chromium.

Step 16. *Confirmation of the presence of zinc.* If you tested for the presence of chromium, use the solution from Step 14. If you were sure chromium was absent, use the solution from Step 12. Make the solution slightly acidic to litmus with 6M HCl added drop by drop. Then add about 3 drops of 0.2M $K_4[Fe(CN)_6]$ and stir. If zinc is present, you obtain a light green precipitate of $K_2Zn_3[Fe(CN)_6]_2$. Centrifuge to make the precipitate more compact for examination.

COMMENTS ON PROCEDURE FOR ANALYSIS
OF GROUP III CATIONS

Step 1. As noted previously, the Group II sulfides are extremely insoluble and precipitate with H_2S in acid solution. The sulfides of the cations in Group III are more soluble than those in Group II and can be precipitated by sulfide ion in a slightly basic solution. An ammonia–ammonium chloride buffer is typically used for controlling pH; since in the precipitating medium the concentrations of hydroxide and sulfide ions are low, about 10^{-5}M and 10^{-6}M, respectively, hydroxide and sulfide complexes do not form, and Group IV cations, if present, do not precipitate. Because of the extreme insolubility of $Al(OH)_3$ and $Cr(OH)_3$, these compounds, rather than their more soluble sulfides, are precipitated.

In the procedure used, the pH is fixed before thioacetamide is added, and some hydroxides may be precipitated, i.e., $Fe(OH)_3$ (rust red), $Fe(OH)_2$ (green), $Cr(OH)_3$ (gray-green), and $Al(OH)_3$ (white). On formation of the sulfides, the color changes. The sulfides have the following colors: FeS (black), CoS (black), MnS (pink), NiS (black), and ZnS (white). Because the sulfide ion reduces iron from the $+3$ state, FeS, rather than $Fe(OH)_3$ or Fe_2S_3, should be the iron precipitate.

On centrifuging out the precipitate, you may find that the remaining liquid is dark, with some fine particles dispersed in it. The NiS does not settle out as completely as a well behaving precipitate should and is responsible for the darkness in the solution. Most of the NiS does precipitate though; no real problem is created, just perhaps a little question in one's mind.

Step 2. The relative insolubilities of CoS and NiS make it possible to separate these sulfides from the others by treatment with 6M HCl. Almost no cobalt(II) dissolves, but some nickel(II) does, and one therefore probably obtains a test for nickel both from the precipitate and from the solution.

Step 4. The use of an alcohol solution of NH_4SCN enhances the stability of the cobalt thiocyanate complex ion and makes the test quite sensitive. If iron is present, a red color tends to be present initially; excess NH_4SCN reagent gives the blue cobalt complex. The formation of yellow potassium hexanitrocobaltate(III), $K_3[Co(NO_2)_6] \cdot 3\,H_2O$, is a less sensitive test for cobalt, but it is also satisfactory. In neither test does nickel interfere.

Step 5. The classic dimethylglyoxime test for nickel is hard to beat. Cobalt does form a colored complex ion with dimethylglyoxime, which must not be confused with a precipitate. However, once you have seen the nickel precipitate made by adding dimethylglyoxime to a dilute nickel solution, you will remember how it looks.

Step 6. This is the crucial step in this analysis. It is necessary that Cr^{3+} be oxidized to CrO_4^{2-} ion. If this is not done, you may miss detecting chromium and in addition may have interference with the confirmatory test for aluminum. After centrifuging, the solution should be distinctly yellow if chromium is present. If it is not yellow, either chromium is absent or you failed in your attempt at oxidation. If things are going well, and you have a fresh solution of H_2O_2, you will probably see oxygen gas evolving as you add the NaOH. The $Cr(OH)_3$ tends to get entrapped in the $Fe(OH)_3$ being formed, and this makes the oxidation harder to accomplish. If the solution after centrifuging is not yellow, it is probably advisable to add 1 ml of H_2O_2 and 1 ml of 6M NaOH, stir it for a minute, boil it for a couple of minutes, and then centrifuge it and proceed.

The oxidation of iron to the +3 state and that of manganese to form MnO_2 proceed without difficulty. The excess hydroxide is needed to form the hydroxide complex ions of zinc and aluminum and to ensure that those species are in the solution phase.

Step 7. Of the three solids possibly present, only the MnO_2 is so weakly basic that it resists solution in acid; some of the manganese does dissolve, but the bulk remains as a brown solid.

Step 8. In acid solution H_2O_2 reduces many higher oxides, including MnO_2, forming more basic oxides; in this case, MnO, or $Mn(OH)_2$, is produced and dissolves very easily in acid.

Step 9. This is the classic test for the presence of manganese. The only interference is by chlorides and other species that can be oxidized, but in excess bismuthate even these finally are all oxidized and the deep purple MnO_4^- ion forms.

Step 10. The test for Fe^{3+} is another one that is hard to miss. Nitric acid also gives a red coloration, which is the reason H_2SO_4 is used in dissolving $Fe(OH)_3$.

Step 11. We have always obtained a confirmatory reaction for nickel at this point when nickel was present. It has tended to be somewhat weaker than that in Step 5, however.

Steps 12 and 13. Students often miss the test for aluminum. If you add the NH_3 slowly when you first precipitate $Al(OH)_3$, you will probably see the precipitate begin to form and then redissolve on mixing. It is not a heavy precipitate but a very characteristic one. Chromium, if present and not oxidized, may interfere, because it forms a similar kind of precipitate. It does not, however, form a lake with aluminon. Add only 2 drops of aluminon reagent, and, if aluminum is present, all the red color is in the precipitate, and the solution is essentially colorless.

Steps 14 and 15. Barium chromate sometimes precipitates slowly, as a fine powder, and there may not be much of it. However, a great deal is not needed for the confirmatory test. The test is more sensitive if you add 1 ml of ether before you put in the H_2O_2. Then shake the mixture. The blue species observed in this test, which is thought to be CrO_5, is soluble in ether and is preferentially extracted into it. Our experience is that ether is not necessary if a noticeable amount of $BaCrO_4$ precipitated.

Step 16. Potassium zinc hexacyanoferrate(II), $K_2Zn_3[Fe(CN)_6]_2$, is nearly white when pure. In this test it is usually light green or blue-green because of contamination with a trace of iron. Under the acidic conditions that prevail, other cations likely to be present do not interfere.

Ions possibly present:

Outline of Procedure for Analysis of Group III Cations

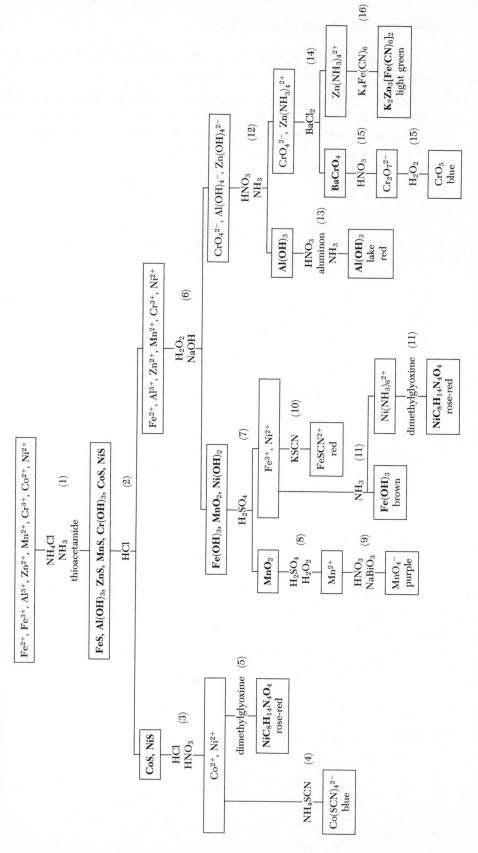

PROBLEMS

1. Write balanced net ionic equations to explain the following observations:
   a. A reddish precipitate forms when solutions of $Fe(NO_3)_3$ and $NaOH$ are mixed.
   b. When $NaOH$ is added to a mixed precipitate of $Al(OH)_3$ and $Fe(OH)_3$, part of the precipitate dissolves.
   c. A white precipitate is formed when ammonia is added to a solution containing Al^{3+}; this precipitate dissolves when the solution is acidified (two reactions).
   d. Addition of $NaOH$ to a solution containing Zn^{2+} gives a white precipitate that dissolves in excess $NaOH$ (two reactions).
   e. A solution of Na_2CrO_4 turns bright orange when it is made strongly acidic.
   f. A pink solution of $CoCl_2$ turns blue on boiling or on addition of HCl.

2. Write balanced net ionic equations for the following oxidation-reduction reactions:
   a. Hydrogen sulfide is added to a solution of $Fe(NO_3)_3$.
   b. Hydrogen peroxide is added to an acidic solution of an iron(II) salt.
   c. The Mn^{2+} ion is oxidized to MnO_2 by $KClO_3$ in acidic solution (assuming reduction of ClO_3^- to Cl^-).
   d. The $Cr_2O_7^{2-}$ complex ion is reduced to Cr^{3+} by H_2S in acidic solution.
   e. The compound $Co(OH)_2$ is oxidized to $Co(OH)_3$ upon standing in contact with air.

3. All the following species can participate in oxidation-reduction reactions. Which of them, in such reactions, can only be reduced? Which can only be oxidized? Which can be either oxidized or reduced, depending upon the conditions?
   a. Hg^{2+} d. H_2O_2 g. Sn^{2+}
   b. $Cu(s)$ e. HCl h. MnO_4^-
   c. Co^{3+} f. $Cr_2O_7^{2-}$ i. Fe^{2+}

4. Which of the sulfides of the ions in Groups II and III precipitate in acidic solution? Which precipitate in basic solution?

5. Describe briefly how you would distinguish between the following:
   a. Cu^{2+} and Zn^{2+} e. $PbS(s)$ and $ZnS(s)$
   b. Cr^{3+} and CrO_4^{2-} f. $AgCl(s)$ and $CoCl_2(s)$
   c. Al^{3+} and Fe^{3+} g. $Fe(OH)_2(s)$ and $Ni(OH)_2(s)$
   d. Mn^{2+} and Zn^{2+} h. $Al_2S_3(s)$ and $NiS(s)$

6. Indicate how you would accomplish the following conversions:
   a. $Cr_2O_7^{2-} \longrightarrow CrO_4^{2-}$ e. $Zn(s) \longrightarrow Zn(NH_3)_4^{2+}$
   b. $Al(s) \longrightarrow Al^{3+}$ f. $Fe^{3+} \longrightarrow FeS(s)$
   c. $Fe^{2+} \longrightarrow Fe^{3+}$ g. $Cr^{3+} \longrightarrow Cr(OH)_4^-$
   d. $Mn^{2+} \longrightarrow MnO_2(s)$ h. $Al_2S_3(s) \longrightarrow Al(OH)_3(s)$

7. Describe exactly what would happen in each of the steps of the Group III analysis with an unknown containing only the following:
   a. Fe^{2+}, Al^{3+}, Mn^{2+} b. Al^{3+}, Cr^{3+}, Co^{2+}

8. Suggest likely explanations for the following mistakes made by students in analyzing Group III unknowns:
   a. Although present, Cr^{3+} is not reported.
   b. Although present, Mn^{2+} is not reported.

9. You are told that a Group III unknown can contain only the following cations. Devise an abbreviated scheme of analysis for these ions, omitting any extra steps.
   a. Al^{3+}, Cr^{3+}, Zn^{2+} b. Co^{2+}, Al^{3+}, Fe^{2+}

10. Describe briefly how you would analyze solutions containing no cations other than the following:
    a. Ag^+, Bi^{3+}, Cr^{3+} b. Cu^{2+}, Sn^{2+}, Al^{3+}

11. A Group III unknown is analyzed with the following results. State which ions are present, which are absent, and which are questionable.
    The precipitate obtained when H_2S is added in basic solution is partially soluble in HCl. Treatment of the HCl solution with NaOH and H_2O_2 gives a strongly colored solution and a precipitate that is partially soluble in H_2SO_4.

12. Indicate how you would determine whether the ions listed as questionable in Problem 11 are present.

13. An unknown that may contain any of the ions in Groups II and III is analyzed with the following results. Indicate, for each of the ions in these two groups, whether it is present, absent, or questionable.
    Addition of H_2S in acidic solution gives a precipitate that is partially soluble in NaOH. The precipitate remaining is dissolved in HNO_3 and treated with NH_3 to give a colored solution and no precipitate. On acidification with HCl, the NaOH solution gives a precipitate that is completely insoluble in hot HCl.
    Treatment of the solution remaining after removal of Group II ions with H_2S in basic solution gives a pure white precipitate.

14. With reference to Problem 13, write net ionic equations involving ions known to be present.

15. A solution may contain any of the ions Ag^+, Hg_2^{2+}, Co^{2+}, and Cr^{3+}. Treatment with HCl gives a white precipitate that turns black upon addition of NH_3. The HCl solution gives a black precipitate when made weakly basic and saturated with H_2S. This precipitate is completely insoluble in HCl.
    Which ions are present? Which are absent? Which are questionable?

16. A solid unknown may contain any of the hydroxides $Pb(OH)_2$, $Cu(OH)_2$, $Fe(OH)_3$, $Al(OH)_3$, and $Zn(OH)_2$. The solid is completely soluble in HCl, partially soluble in NH_3, and essentially insoluble in NaOH. On the basis of this information, identify the solid as completely as possible.

101. Write balanced net ionic equations for the reactions that occur when
     a. Hydrogen sulfide is added to a solution of an iron(II) salt.
     b. Solutions of $FeCl_3$ and KSCN are mixed.
     c. A basic solution containing S^{2-} ions is added to a solution of $Al(NO_3)_3$.
     d. Hydrogen sulfide in strongly acidic solution is added to a solution containing Cu^{2+} and Zn^{2+}.
     e. Ammonia is added to $Zn(OH)_2(s)$.
     f. A strong acid is added to barium chromate.
     g. Ammonia is added to a solution of a Ni^{2+} salt (two reactions).

102. Complete and balance the following oxidation-reduction equations:
     a. $Fe(OH)_2(s) + O_2(g) \longrightarrow Fe(OH)_3(s)$
     b. $Fe^{3+} + I^- \longrightarrow$
     c. $Zn(s) + H^+ \longrightarrow$
     d. $Mn(OH)_2(s) + Br_2 \longrightarrow MnO_2(s)$ (basic solution)
     e. $Cr^{3+} + H_2O_2 \longrightarrow CrO_4^{2-}$ (basic solution)
     f. $Co^{2+} + O_2(g) + NH_3 \longrightarrow Co(NH_3)_6^{3+}$ (basic solution)

103. Which of the following species, in oxidation-reduction reactions, can act only as an oxidizing agent? Which can act only as a reducing agent? Which can act as either an oxidizing or a reducing agent?
     a. Hg_2^{2+} d. H_2S g. Sn^{4+}
     b. Cu^{2+} e. HNO_3 h. $Mn(s)$
     c. Co^{2+} f. CrO_4^{2-} i. Fe^{3+}

104. Which of the hydroxides of the ions in Groups II and III are soluble in excess NH_3? Which are soluble in excess NaOH?

105. Describe briefly how you would distinguish between the following:
     a. Fe^{2+} and Fe^{3+} e. $Fe(OH)_3(s)$ and $Al(OH)_3(s)$
     b. Fe^{2+} and Cu^{2+} f. $CuS(s)$ and $FeS(s)$
     c. $Zn(NH_3)_4^{2+}$ and $Zn(OH)_4^{2-}$ g. $SnS(s)$ and $NiS(s)$
     d. $Co(SCN)_4^{2-}$ and $Fe(SCN)^{2+}$ h. $AgNO_3(s)$ and $Zn(NO_3)_2(s)$

106. Indicate how you would accomplish the following conversions:
     a. $Zn(s) \longrightarrow ZnCl_2(s)$ e. $Zn(OH)_4^{2-} \longrightarrow Zn(NH_3)_4^{2+}$
     b. $Co^{2+} \longrightarrow Co^{3+}$ f. $Fe(OH)_2(s) \longrightarrow Fe(OH)_3(s)$
     c. $Fe^{3+} \longrightarrow Fe^{2+}$ g. $Cr^{3+} \longrightarrow Cr_2O_7^{2-}$
     d. $Mn^{2+} \longrightarrow MnO_4^-$ h. $Fe^{3+} \longrightarrow Fe(CN)_6^{3-}$

107. Describe exactly what would happen in each step of the Group III analysis with an unknown containing only the following:
     a. Cr^{3+}, Ni^{2+}, Fe^{3+} b. Co^{2+}, Al^{3+}, Mn^{2+}

108. The following errors are frequently made by students in analyzing Group III unknowns. Suggest a likely explanation for each error.
     a. Although not reported, Al^{3+} is present.
     b. Although reported, Al^{3+} is not present.

109. Assume that a Group III unknown cannot contain any cations other than the following. Devise an abbreviated scheme of analysis for these ions, omitting all extra steps.
     a. Fe^{2+}, Ni^{2+}, Al^{3+} b. Fe^{2+}, Mn^{2+}, Zn^{2+}

110. Devise abbreviated schemes of analysis for unknowns containing no cations other than the following:
     a. Ag^+, Cd^{2+}, Cu^{2+}, Cr^{3+} b. Pb^{2+}, Sb^{3+}, Co^{2+}, Zn^{2+}

111. A Group III unknown exhibits the following behavior. State which ions are present, which are absent, and which are questionable.

    Addition of NH_3 and NH_4Cl to the unknown gives a reddish precipitate that turns black when H_2S is added. The final precipitate is completely soluble in HCl. Treatment of the HCl solution with H_2O_2 and NaOH gives a colorless solution and a precipitate that is completely soluble in H_2SO_4.

112. With reference to Problem 111, indicate how you would:
    a. Determine which of the ions listed as questionable are present.
    b. Confirm the ions listed as present.

113. An unknown that may contain any of the ions in Group I, II, and III exhibits the following behavior. Indicate whether each ion is present, absent, or questionable.

    Treatment of the unknown with HCl give a white precipitate that is completely soluble in NH_3. The HCl solution does not give a precipitate when treated with H_2S. However, when the solution is made weakly basic with NH_3, a precipitate forms. This precipitate is partially soluble in HCl, leaving a black residue. Treatment of the HCl solution with H_2O_2 and NaOH gives a precipitate that completely dissolves in H_2SO_4. This solution gives a negative test result when treated with dimethylglyoxime.

114. With reference to Problem 113, write net ionic equations to describe the reactions undergone by the ions known to be present.

115. An unknown may contain any of the ions Pb^{2+}, Cu^{2+}, Al^{3+}, and Ni^{2+}. Treatment of the unknown with H_2S in acidic solution gives a black precipitate. The precipitate is dissolved in nitric acid and treated with ammonia to give a deeply colored solution and a white precipitate. The solution remaining from the H_2S precipitation is boiled to remove H_2S and is then treated with excess NaOH. A colored precipitate forms; upon careful acidification and treatment with NH_3, the solution gives no precipitate.

    Which ions are present? Which are absent? Which are questionable?

116. A solid unknown may contain any of the sulfides PbS, Bi_2S_3, FeS, NiS, and Al_2S_3. The solid is completely insoluble in concentrated NaOH. It is partially soluble in cold, dilute HCl. Following the treatment with HCl, the precipitate is brought into solution with HNO_3. When this solution is treated with NH_3, no precipitate forms.

    On the basis of this information, identify the solid as completely as possible.

LABORATORY ASSIGNMENTS

1. Using a known sample containing each of the Group III cations, analyze the solution by the procedure given in the text. Compare your observations with those described. Obtain a Group III unknown and analyze it by the procedure. Report your observations and conclusions on a Group III flow chart.

2. You will be furnished an unknown containing only Group III cations selected from the following sets. When you have been assigned a set, prepare a procedure by which you can establish the presence of those cations in your unknown. Test the procedure to see that it works and draw a flow chart showing all the reagents to be used in the various steps and the species that may be formed. Obtain an unknown from your instructor and analyze it by your procedure. Report your

observations and conclusions on your flow chart. The sets of Group III cations to be considered are:

a. Ni^{2+}, Al^{3+}, Cr^{3+}
b. Co^{2+}, Mn^{2+}, Zn^{2+}
c. Fe^{3+}, Co^{2+}, Cr^{3+}
d. Al^{3+}, Zn^{2+}, Ni^{2+}
e. Cr^{3+}, Mn^{2+}, Al^{3+}
f. Al^{3+}, Co^{2+}, Ni^{2+}
g. Fe^{3+}, Zn^{2+}, Mn^{2+}
h. Ni^{2+}, Co^{2+}, Zn^{2+}

3. You will be given an unknown containing only four cations selected from Groups I, II, and III, with at least one cation chosen from each. Develop as simple a procedure as you can for analyzing such a limited unknown. Draw a flow diagram for your procedure, indicating the reagents to be used in each step and the observations to be anticipated. Show your procedure to your instructor for approval, and then analyze an unknown by your method. When you believe you know which four cations are present, make up a known solution containing those ions and test it to see that it behaves in the same way as your unknown. Indicate on your flow chart your observations and conclusions on the unknown, and turn the completed chart in to your instructor.

4. Your instructor will assign to you in advance an unknown that may contain only the cations indicated in one of the following sets. Before coming to laboratory, develop a procedure by which you can analyze the unknown for the ions in your set. Do not include any unnecessary steps, and try to make your scheme for separation as simple as possible. Draw a complete flow chart for your scheme, indicating the reagents to be used in each step, the species that may be formed in each step, and the observations that will confirm the presence of each cation. Test your procedure with a sample containing each of the cations in your set, modifying your approach as necessary. When you are sure that your scheme of analysis works, use it to determine the cations present in an unknown furnished by your instructor. Record on your flow chart your observations and conclusions for the unknown. Your set of cations will be one of the following:

a. Al^{3+}, Fe^{3+}, Zn^{2+}, Cd^{2+}, and Pb^{2+}
b. Fe^{3+}, Ni^{2+}, Cr^{3+}, Bi^{3+}, and Hg^{2+}
c. Al^{3+}, Ni^{2+}, Cr^{3+}, Sn^{4+}, and Hg^{2+}
d. Ni^{2+}, Zn^{2+}, Mn^{2+}, Cu^{2+}, and Ag^{+}
e. Co^{2+}, Zn^{2+}, Mn^{2+}, Cd^{2+}, and Sb^{3+}
f. Zn^{2+}, Mn^{2+}, Cu^{2+}, Ag^{+}, and Pb^{2+}
g. Al^{3+}, Ni^{2+}, Co^{2+}, Cd^{2+}, and Sn^{4+}
h. Fe^{3+}, Co^{2+}, Zn^{2+}, Hg^{2+}, and Pb^{2+}

SECTION 4

THE PROPERTIES OF THE CATIONS IN GROUP IV— Ba^{2+}, Ca^{2+}, Mg^{2+}, Na^+, K^+, NH_4^+

BARIUM, Ba^{2+}

Most of the common barium salts are soluble in water and in dilute strong acids. The main exception is the sulfate, $BaSO_4$, which thereby becomes useful in analysis for barium. The salts of barium are typically white and colorless in solution. The hydroxide is a strong base and is moderately soluble, about 0.2M at 25° C. There are two oxygen compounds of barium, BaO and BaO_2, the latter being a peroxide that liberates H_2O_2 on solution in water. Many of the salts of barium crystallize as hydrates; the common sources of barium ion in solution are $BaCl_2 \cdot 2\,H_2O$ and $Ba(NO_3)_2$. Barium does not readily form complex ions.

CHARACTERISTIC REACTIONS OF THE BARIUM ION

1. **Sodium Hydroxide and Ammonia.** These reagents do not react with solutions containing Ba^{2+}.

2. **Soluble Sulfides.** Barium sulfide cannot be precipitated by sulfide ion in acidic or basic solution. When BaS dissolves in water, a very strongly basic solution is obtained because of hydrolysis of the sulfide ion. The hydroxide, however, does not usually precipitate, because it is relatively soluble.

3. **Sodium Carbonate.** On addition to solutions containing barium ion, this reagent causes $BaCO_3$ to form as a white precipitate. The $BaCO_3$ is very soluble in dilute acid solution, and even dissolves in dilute acetic acid; it is not soluble in strong bases and in ammonia.

4. **Soluble Sulfates.** Sulfuric acid or any soluble sulfate causes precipitation of $BaSO_4$, a very finely divided, heavy, white powder that is essentially insoluble in all common reagents. If $BaSO_4$ is heated with a concentrated solution of Na_2CO_3, the following reaction occurs to some extent:

$$BaSO_4(s) + CO_3^{2-} \rightleftharpoons BaCO_3(s) + SO_4^{2-}$$

The reaction goes more nearly completely in molten Na_2CO_3. In either approach, separation of the carbonate, which is soluble in acid, allows one to bring Ba^{2+} back into solution.

5. **Potassium Chromate.** A finely divided yellow precipitate of $BaCrO_4$ forms on addition of this reagent. The $BaCrO_4$ is soluble in dilute acids but not in bases:

$$2\,BaCrO_4(s) + 2\,H^+ \rightleftharpoons 2\,Ba^{2+} + Cr_2O_7^{2-} + H_2O$$

6. **Sodium Oxalate.** White barium oxalate, $BaC_2O_4 \cdot H_2O$, precipitates on addition of soluble oxalates. Barium oxalate is slightly soluble in water and dissolves in boiling dilute acetic acid as well as dilute solutions of strong acids:

$$BaC_2O_4(s) + HC_2H_3O_2 \rightleftharpoons Ba^{2+} + HC_2O_4^- + C_2H_3O_2^-$$

7. **Sodium Monohydrogen Phosphate.** White, flocculent $BaHPO_4$ forms in neutral solutions containing HPO_4^- ion. In more basic solutions, insoluble $Ba_3(PO_4)_2$ forms. Both substances are soluble in dilute acids at a pH of 3 or less:

$$BaHPO_4(s) + H^+ \rightleftharpoons Ba^{2+} + H_2PO_4^-$$

8. **Flame Test.** Barium salts give a yellow-green color to a Bunsen flame.

CALCIUM, Ca^{2+}

Calcium salts, like those of barium, are typically soluble in water and in dilute acids. The hydroxide is somewhat less soluble than that of barium, the saturated solution being about 0.02M; it too behaves as a strong base. Calcium salts, for the most part, are white and often crystallize as hydrates. Calcium seldom forms complex ions.

CHARACTERISTIC REACTIONS OF THE CALCIUM ION

1. **Sodium Hydroxide.** The hydroxide $Ca(OH)_2$ can be precipitated from solution by addition of 6M NaOH if Ca^{2+} is present in moderate concentration. The hydroxide is soluble in very dilute acid.

2. **Ammonia.** There is no reaction with this reagent.

3. **Soluble Sulfides.** The Ca^{2+} ion does not react with this reagent.

4. **Sodium Carbonate.** White $CaCO_3$ precipitates. It filters more readily on being heated, and it is soluble in dilute acids.

5. **Sodium Oxalate.** This reagent produces a highly insoluble white precipitate of $CaC_2O_4 \cdot H_2O$ from solutions containing Ca^{2+} ion. This oxalate does not dissolve appreciably in acetic acid, but it does go into solution in dilute strong acids:

$$CaC_2O_4(s) + H^+ \rightleftharpoons Ca^{2+} + HC_2O_4^-$$

6. **Sodium Monohydrogen Phosphate.** A white precipitate that is soluble in acids forms on addition of this reagent.

7. **Flame Test.** Calcium salts have a yellow-red color in a Bunsen flame.

MAGNESIUM, Mg^{2+}

Magnesium salts are soluble in water and in dilute acids. They are all white and very frequently crystallize as hydrates. Magnesium forms few complex ions. The hydroxide is much less soluble than are the hydroxides of barium and calcium. Other magnesium compounds tend to be somewhat more soluble than those of the other alkaline earths.

CHARACTERISTIC REACTIONS OF THE MAGNESIUM ION

1. **Sodium Hydroxide and Ammonia.** White gelatinous $Mg(OH)_2$ precipitates in the presence of these reagents. This hydroxide goes into solution in dilute acids and even dissolves in ammonium salts because of their very slight acidity.

2. **Soluble Sulfides.** The Mg^{2+} ion does not react in acid and in neutral media, or in the presence of ammonium salts. In moderately basic solution, $Mg(OH)_2$ precipitates.

3. **Sodium Carbonate.** A white basic salt (possible formula $Mg_4(CO_3)_3(OH)_2$) precipitates from alkaline solutions of carbonate ion. This salt dissolves very readily in acids.

4. **Soluble Sulfates.** There is no reaction with these reagents.

5. **Potassium Chromate.** There is no reaction.

6. **Sodium Oxalate.** If the concentration of Mg^{2+} is moderately high, a precipitate of $MgC_2O_4 \cdot 2\,H_2O$ may form. This salt is soluble in very dilute acids and even to some extent in ammonium salts.

7. **Sodium Monohydrogen Phosphate.** In neutral solution a light precipitate of $MgHPO_4 \cdot 7\,H_2O$ is obtained. In an ammonia–ammonium chloride buffer a characteristic crystalline precipitate is produced:

$$Mg^{2+} + NH_3 + HPO_4^{2-} \rightleftharpoons NH_4MgPO_4(s)$$

The precipitate dissolves readily in very dilute acids, but it is essentially insoluble in ammonia and in more alkaline solutions.

8. **Magnesium Reagent.** The hydroxide $Mg(OH)_2$ forms a characteristic blue lake with a dilute solution of 4-(p-nitrophenylazo)resorcinol, also called magnesium reagent. The reagent is added to a slightly acidic solution of Mg^{2+}; NaOH is then added slowly until the hydroxide precipitate forms. Nickel and cobalt hydroxides also form blue lakes with this reagent.

AMMONIUM, NH_4^+

Although ammonium ion is not a metallic cation, it forms salts with properties similar to those of the alkali metals and is usually included in schemes of qualitative analysis. Ammonium salts are all white and soluble and, if their anions are those of strong acids, are slightly acidic in solution because of partial dissociation of NH_4^+:

$$NH_4^+ \rightleftharpoons NH_3 + H^+$$

Ammonium ion is produced by the reverse of the preceding reaction when acids are added to ammonia solutions.

CHARACTERISTIC REACTIONS OF THE AMMONIUM ION

1. **Sodium of Potassium Hydroxide.** Concentrated solutions of hydroxide ion cause formation and evolution of NH_3 gas when added to solutions containing NH_4^+:

$$NH_4^+ + OH^- \rightleftharpoons NH_3(g) + H_2O$$

The gas can be detected by its odor or by its ability to turn moistened red litmus paper blue.

2. **Heat.** Ammonium salts tend to be much more volatile and unstable to heat than are those of the alkali metals. The products are often NH_3 and an acid, but nitrogen gas or nitric oxide may be formed, depending on

the salt. In any event, heating an ammonium salt to red heat results in its being volatilized, and consequently removed from the system.

SODIUM, Na+

Salts of sodium are typically soluble and white. Sodium ion in water solution is chemically quite inert, and like the other cations in Group IV it cannot be chemically reduced to the metal. Sodium hydroxide is very soluble and the common source of high concentrations of OH^- ion in aqueous solutions. Sodium very rarely forms complex ions. Its salts frequently crystallize as hydrates.

CHARACTERISTIC REACTIONS OF THE SODIUM ION

1. **Flame Test.** In a Bunsen flame sodium salts give off a very strong, characteristic yellow light. Even traces of Na^+ are active in this regard; therefore the duration and intensity of the emitted light must be considered when testing for the presence of sodium.

2. **Precipitating Reagents.** There are no really insoluble, easily prepared compounds of sodium; therefore, sodium analysis is ordinarily done by physical methods. Both the qualitative and the quantitative analytical methods ordinarily depend on the observation of the emission spectrum of sodium. One of the least soluble compounds of sodium is the zinc uranyl acetate, $NaZn(UO_2)_3(C_2H_3O_2)_9$, a pale yellow substance that can be precipitated from moderately concentrated solutions of Na^+ ion.

POTASSIUM, K+

Salts of potassium are similar to those of sodium in their general properties. They are nearly all white and soluble in water. Potassium hydroxide is very soluble and is a strong base.

CHARACTERISTIC REACTIONS OF THE POTASSIUM ION

1. **Flame test.** Potassium salts, preferably the chloride or nitrate, emit a violet light in the Bunsen flame. The test is much less sensitive than that for sodium. Sodium interferes, and if it is present, blue glass is placed in front

of the flame to absorb the sodium radiation and allow the observation of that from potassium.

2. **Precipitating Reagents.** There are no highly insoluble, easily prepared, compounds of potassium. Hence, as with sodium, both qualitative and quantitative analyses of potassium are ordinarily made by physical measurements. Among the least soluble of the potassium salts are yellow potassium sodium hexanitrocobaltate(III), $K_2Na[Co(NO_2)_6]$, and yellow potassium hexachloroplatinate(IV), $K_2[PtCl_6]$.

3. **Oxidizing and Reducing Agents.** Potassium metal, like all the metals obtainable from the Group IV cations, is a very strong reducing agent. Both potassium and sodium metal react violently with water, producing hydrogen gas and the metal hydroxide. Barium and calcium react more slowly with water, whereas magnesium reacts only under slightly acidic conditions.

GENERAL DISCUSSION OF PROCEDURE FOR ANALYSIS OF THE GROUP IV CATIONS Ba^{2+}, Ca^{2+}, Mg^{2+}, Na^+, K^+, NH_4^+

The ions remaining in solution after the separation of the Group III cations are those of the alkaline earths, the alkali metals, and ammonium ion. These ions pass through the analysis scheme unaffected by chloride ion or sulfide ion in acidic and in basic media.

Because sodium ion is a common impurity in chemical reagents and because ammonium salts are used as reagents in the general separation scheme, tests for Na^+ and NH_4^+ ion are carried out on an original sample of solution, one that has not been treated with any reagents. The characteristic flame test is used to detect the presence of sodium; the evolution of NH_3 gas from a sample made basic with NaOH is used for identification of ammonium ion. Potassium ion is also identified by flame test of the original sample. Precipitation of salts of Na^+ or K^+ from the solution containing Group IV cations is not practical.

Barium and calcium are separated from the alkali metal ions and, for the most part, from magnesium by precipitation of their carbonates under the slightly basic conditions prevailing in an ammonia–ammonium chloride buffer. The precipitate is dissolved in acetic acid and successively, barium is separated as the chromate, calcium as the oxalate, and magnesium as an ammonium phosphate salt, in that order. Magnesium and potassium are also tested for in the solution remaining from the carbonate precipitation.

When performed on a sample containing only Group IV cations, the procedure we use works very nicely. In the case of a general unknown, from which cations in Groups I, II, and III have been precipitated, one may encounter less definitive results, because dilution and coprecipitation effects associated with the group separations tend to remove the Group IV cations from the system. Also the ammonium salts that have been added have a marked diluting effect on the alkaline earth cations. To remedy this situation, at least partially, ammonium salts are removed from the sample of general unknown by sublimation before precipitation of the carbonates is attempted, and tests on the most soluble cations are carried out on original, untreated samples.

PROCEDURE FOR ANALYSIS OF GROUP IV CATIONS

Step 1. If you are working with only the Group IV cations, you may assume, unless told otherwise, that 5 ml of sample contains the equivalent of 1 ml of 0.1M solutions of the nitrate or chloride salts of one or more of those cations. Pour 5 ml of your sample into a 30 ml beaker and boil the solution down to a volume of 2 ml. Then add 0.5 ml of 6M HCl, mix the solution, and transfer it to a test tube.

If you are analyzing a general unknown, prepare for Group IV analysis by transferring the solution remaining after removal of the Group III cations to a 30 ml beaker and boiling it down to a volume of 2 ml. Transfer the liquid to a test tube and centrifuge out any solid matter, which you may discard. Put the liquid back into the beaker, add 1 ml of 6M HCl, and proceed to boil the material essentially to dryness. Transfer the beaker to a hood, and carefully heat the dry solid to drive off all the ammonium salts that were added in previous steps. Stop heating the solid when the visible smoke from these salts is no longer being evolved. Let the beaker cool, and then add 0.5 ml of 6M HCl and 2 ml of water. Warm the solution gently to dissolve the remaining salts; discard any insoluble material after centrifuging it out. Pour the solution into a test tube.

Step 2. To your Group IV sample prepared in Step 1, add 6M NH_3 until the solution becomes basic. Then add 1 ml of 1M $(NH_4)_2CO_3$ and stir. Put the test tube in the hot, *but not boiling,* water bath, and without any heating, leave the tube in the bath for two minutes, stirring occasionally. Centrifuge and decant the liquid, which may contain K^+ and Mg^{2+}, into a test tube. Wash the precipitate, which may contain $BaCO_3$, $CaCO_3$, and possibly some $Mg(OH)_2$ or $MgCO_3$, with 2 ml of water. Centrifuge out the precipitate, and discard the wash.

Step 3. *Confirmation of the presence of barium.* Add 0.5 ml of 6M acetic acid to the precipitate from Step 2 and stir to dissolve the solid. Add 1 ml of water and mix; then add 0.5 ml of 1M K_2CrO_4. A yellow precipitate indicates the presence of barium. Stir for a minute and centrifuge out the solid. Decant the liquid, which may contain Ca^{2+} and Mg^{2+}, into a test tube. Wash the solid with 2 ml of water; centrifuge, and discard the wash. Dissolve the solid in 0.5 ml of 6M HCl; add 1 ml of water. Then add 0.5 ml of 6M H_2SO_4, and stir the solution for 30 seconds. A white precipitate of $BaSO_4$ establishes the presence of barium. Centrifuge to separate the white solid from the orange solution.

Step 4. *Confirmation of the presence of calcium.* Add 6M NH_3 to the solution from Step 3 until it becomes basic; at this point the solution becomes yellow. Add 0.5 ml of 1M $K_2C_2O_4$, stir, and let the solution stand for a minute.

A white precipitate of $CaC_2O_4 \cdot H_2O$ is confirmation for the presence of calcium. Centrifuge out the white solid, and pour the liquid, which may contain Mg^{2+}, into a test tube. Wash the solid with 2 ml of water; centrifuge, and discard the wash.

Dissolve the solid in 1 drop of 6M HCl. Perform a flame test on the resulting solution. A fleeting orange-red sparkly flame is due to calcium. This color appears before the yellow sodium flame, and it is best observed if you have a small drop of the solution on the test loop; the color lasts only a fraction of a second. Compare your observations here with those obtained with a 0.1M $CaCl_2$ solution.

Step 5. *Confirmation of the presence of magnesium.* To the solution from Step 4, add 0.5 ml of 1M Na_2HPO_4. Stir the solution, warm it gently, and let it stand for a minute. A white precipitate is highly indicative of the presence of magnesium. Centrifuge out the solid, and discard the liquid. Wash the solid with 2 ml of water; centrifuge, and discard the wash. Dissolve the solid in a few drops of 6M HCl. Add 1 ml of water and then 2 or 3 drops of magnesium reagent, 4-(p-nitrophenylazo)resorcinol. Stir the solution, and then, drop by drop, add 6M NaOH until the solution is basic. If magnesium is present, a medium blue precipitate of $Mg(OH)_2$ with adsorbed magnesium reagent forms. Centrifuge out the precipitate from the almost colorless solution for better observation of the solid.

Step 6. *Alternative confirmation of the presence of magnesium.* Repeat Step 5 on one half of the solution obtained in Step 2. If magnesium is present in your sample, you will probably detect it in both Step 5 and Step 6; we have found that the amount of blue precipitate is usually greater in Step 6.

Step 7. *Confirmation of the presence of sodium and potassium.* Perform a flame test on 1 ml of the *original* sample, using a portion that has not been treated in any way. A strong yellow flame, which persists for one or two seconds, is confirmatory evidence for the presence of sodium. Since the test is *very* sensitive to traces of sodium, compare the intensity and duration of the flame you obtain with that from a sample of distilled water and that from 0.1M NaCl solution.

We also perform a flame test on the original sample to detect potassium. The potassium flame is violet and usually lasts for only a few moments, perhaps $\frac{1}{2}$ second. The test is much less sensitive than that for sodium, and is best observed with crystals obtained by evaporating 1 ml of sample to near dryness. Look for the potassium flame through one or two thicknesses of blue cobalt glass, which absorbs any sodium emission. The flame from potassium looks reddish-violet through the glass. Compare the flame you observe with that from saturated KCl solution.

Since potassium is much less likely to be a contaminant in your reagents than is sodium, it is also possible to perform a meaningful test for potassium on the rest of the solution from Step 2. Boil that solution down to a volume of only a few drops, and carry out the test for potassium on those drops.

Because this solution may be quite concentrated in K^+, the test for potassium should be easy to see. In addition, other cations that might interfere with the potassium test in the original sample are not present in the solution remaining after the Group IV precipitation.

Step 8. *Confirmation of the presence of NH_4^+ ion.* Pour the 1 ml of *original* sample used in the flame test for Na^+, or 1 ml of a fresh sample, into a 30 ml beaker. Moisten a piece of red litmus paper and put it on the bottom of a small watch glass. Add 1 ml of 6M NaOH to the sample in the beaker and swirl to stir. Cover the beaker with the watch glass, and then gently heat the solution to the boiling point; do not boil it, and be careful that no liquid solution comes in contact with the litmus paper. If ammonium ion is present, the litmus paper gradually turns blue as it is exposed to the evolved vapors of NH_3. Remove the watch glass and try to smell the ammonia. If only a small amount of ammonium ion is present, the smell test will probably not detect it, whereas the litmus test will.

COMMENTS ON PROCEDURE FOR ANALYSIS OF GROUP IV CATIONS

Step 1. If you have been working with a general unknown, by the time you get to Group IV analysis you have added a substantial number of reagents to the original sample. In particular, ammonium salts have been added; they dilute the Group IV cations and make their precipitation and detection more difficult. Since all ammonium salts are relatively volatile, they are removed from the Group IV metallic cations by simple heating. While subliming the ammonium salts, do not overheat the solid, but keep the temperature at the point at which the salts sublime relatively slowly and smoothly. The nonvolatile salts remaining are easier to dissolve if they have not been heated above the minimal temperature necessary for sublimation.

Step 2. In this step the carbonates of barium, calcium, and possibly the hydroxide or carbonate of magnesium precipitate. The warming of the mixture promotes formation of well defined precipitates. In the ammonia–ammonium chloride buffer used, magnesium hydroxide or carbonate should not precipitate, but Mg^{2+} does tend to coprecipitate with the alkaline earths.

Step 3. Neither calcium nor magnesium chromate precipitates under the slightly acidic conditions that prevail here. However, $BaCrO_4$ is essentially quantitatively removed. Its solubility in HCl and reprecipitation of the Ba^{2+} as barium sulfate is definitive evidence for the presence of barium. Lead would not interfere with this test.

Step 4. Neither magnesium oxalate nor barium oxalate from the residual barium present precipitates at this point. A precipitate on addition of oxalate is really definitive evidence for the presence of calcium. Nevertheless, the flame test gives that added degree of confidence that is so reassuring. The calcium flame test lasts only a moment and typically consists of a few red-orange flashes plus a small amount of flame.

Steps 5 and 6. The blue lake test for magnesium is excellent. There is interference if cobalt or nickel is present, but that is highly unlikely. You will see the blue $Mg(OH)_2$ precipitate gradually forming as you add the NaOH. Considering that this is the last cation to be precipitated in this scheme, the test is remarkably sensitive.

Step 7. It is very difficult to carry out convincing precipitations of the salts of sodium, potassium, and ammonium ions. The flame tests for sodium and potassium are quite adequate, but they must be done with care. Consult Chapter 2 for the proper procedure to use in making a flame test. Be sure that your wire is as clean as you can make it before attempting any flame tests. Because your sample will probably show positive results of a sodium test even if almost no sodium is present, be sure to compare your sample with a standard NaCl solution and with distilled water. The potassium test is more easily observed in a partially darkened room.

Step 8. If only a small amount of ammonium salt is present, you probably cannot smell it. Spattering from the alkaline sample of course turns the litmus paper blue, but that would not give the smooth, gradual change in color that you get from the NH_3 vapors. Obviously, this test must be done on a sample that has not had either ammonia or ammonium salts added to it.

Outline of Procedure for Analysis of Group IV Cations

Ions possibly present:

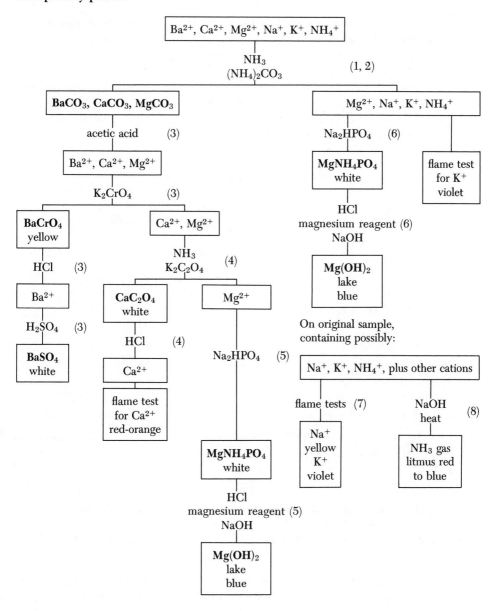

PROBLEMS

1. Write balanced net ionic equations to explain the following observations:
    a. A white precipitate forms when solutions of $BaCl_2$ and Na_2SO_4 are mixed.
    b. Water solutions of Group IV sulfides are basic to litmus.
    c. Barium oxalate is soluble in dilute acid.
    d. Addition of OH^- to a solution of $Mg(NO_3)_2$ gives a precipitate that dissolves when the solution is acidified (two reactions).
    e. Addition of Na_2HPO_4 to an ammoniacal solution of Mg^{2+} gives a precipitate.
    f. A gas with a pungent odor is evolved when concentrated NaOH is added to a solution of NH_4Cl.

2. Describe briefly how you would distinguish between the following:
    a. Ba^{2+} and Ca^{2+} f. $NaCl(s)$ and $BaCl_2(s)$
    b. Mg^{2+} and Ca^{2+} g. $CaC_2O_4(s)$ and $BaC_2O_4(s)$
    c. Mg^{2+} and Na^+ h. $Mg(OH)_2(s)$ and $Ca(OH)_2(s)$
    d. Na^+ and K^+ i. $MgSO_4(s)$ and $BaSO_4(s)$
    e. Na^+ and NH_4^+ j. $NaNO_3(s)$ and $Na_2SO_4(s)$

3. Describe briefly how you would distinguish between the following:
    a. K^+ and Fe^{3+} e. $MgCl_2(s)$ and $MnCl_2(s)$
    b. NH_4^+ and H^+ f. $Mg(OH)_2(s)$ and $Zn(OH)_2(s)$
    c. Cu^{2+} and Ca^{2+} g. $BaCl_2(s)$ and $PbCl_2(s)$
    d. CO_3^{2-} and SO_4^{2-} h. $CaCO_3(s)$ and $CoCO_3(s)$

4. Indicate how you would accomplish the following conversions:
    a. $BaCl_2(s) \longrightarrow BaSO_4(s)$ d. $BaCO_3(s) \longrightarrow BaSO_4(s)$
    b. $NH_4Cl(s) \longrightarrow NH_3(g)$ e. $MgCl_2(s) \longrightarrow Mg(OH)_2(s)$
    c. $NaOH(s) \longrightarrow NaNO_3(s)$ f. $NaI(s) \longrightarrow NaNO_3(s)$

5. Give the formula of a species that brings the cation in each of the following solids into solution and write a net ionic equation to describe the reaction involved.
    a. $BaSO_4$ c. $AgCl$ e. MgC_2O_4 g. $Fe(OH)_3$
    b. $Mg(OH)_2$ d. NiS f. $CaCO_3$ h. SnS

6. Give the formula of a reagent that brings into solution the following:
    a. $BaCO_3$ but not $BaSO_4$ c. Na_2S but not CuS
    b. $AgCl$ but not $PbCl_2$ or $Mg(OH)_2$ d. $BaCrO_4$ but not $PbCrO_4$

7. Explain why:
    a. In working with a general unknown, it is advisable to evaporate the solution to dryness and to heat the material strongly before testing for Group IV ions.
    b. Ammonium ions are added as well as NH_3 in precipitating Group III cations.
    c. The solution in Step 3 of Group IV analysis is made weakly acidic rather than neutral.
    d. Ammonia is added at the beginning of Step 4.
    e. A piece of blue glass is used in the flame test for K^+.

8. Describe exactly what would happen in each step of Group IV analysis with unknowns containing only the following:
    a. Ba^{2+} and Mg^{2+} b. Ca^{2+}, Mg^{2+}, and NH_4^+

9. Work out a simplified scheme of analysis, omitting any extra steps, for an unknown containing only the following:
   a. Ba^{2+} and Mg^{2+} b. Ca^{2+}, K^+, and NH_4^+ c. K^+ and Na^+

10. Describe an efficient system of analysis for unknowns that contain no cations other than the following:
    a. Ag^+, Bi^{3+}, Mn^{2+}, and NH_4^+ b. Ni^{2+}, Al^{3+}, Ca^{2+}, and Na^+

11. A student, in analyzing a Group IV unknown, makes the following observations. On the basis of this information, indicate whether each ion is present, absent, or questionable.

    The unknown is made basic with NH_3, treated with $(NH_4)_2CO_3$, and heated to obtain a white precipitate. This solid is dissolved in acetic acid; addition of K_2CrO_4 to the solution gives a yellow precipitate.

    Another portion of the unknown gives a strong yellow flame in the flame test. When the unknown is made strongly basic with NaOH and heated, a gas is given off which turns red litmus blue.

12. A student analyzes an unknown that he is told may contain only the cations Cu^{2+}, Cd^{2+}, Fe^{3+}, Al^{3+}, Ba^{2+}, Ca^{2+}. Treatment with H_2S in acidic solution gives a black precipitate. The solution is boiled to remove H_2S and made basic with NH_3, giving a white precipitate that is soluble in NaOH. The ammoniacal solution is treated with H_2SO_4 to give a white precipitate.

    Which ions are present? Which are absent? Which are questionable?

13. A student analyzes a general unknown that may contain any of the cations in Groups I, II, III, and IV. He makes the following observations:

    Treatment of the unknown with HCl gives a white precipitate. When the precipitate is treated with NH_3, part of it dissolves; the residue is black.

    Treatment of the HCl solution with H_2S gives a precipitate that is completely soluble in NaOH. Acidification of the NaOH solution gives a precipitate that is completely insoluble in HCl.

    The solution remaining after removal of the Group II ions is made basic with NH_3 and NH_4^+ and saturated with H_2S. A precipitate forms that is partially soluble in HCl. Treatment of the HCl solution with H_2O_2 and NaOH gives a colorless solution and a black precipitate that is partially soluble in H_2SO_4.

    The solution remaining after removal of Group III ions is evaporated to dryness and heated. The residue is dissolved in HCl, treated with NH_3 and $(NH_4)_2CO_3$, and heated to give a white precipitate. This precipitate is dissolved in acetic acid; addition of K_2CrO_4 to the acetic acid solution gives a yellow precipitate. Treatment of the solution with NH_3 and $K_2C_2O_4$ fails to give a precipitate. However, when NaOH is added to make the solution strongly basic, a white precipitate forms.

    Which ions are present? Which are absent? Which are questionable?

14. A solid unknown may contain any of the following compounds: $Cu(NO_3)_2$, $Ni(OH)_2$, FeS, NH_4Cl, and $CaCO_3$. The unknown is extracted with water to give a colorless solution that is weakly acidic. The solid remaining is treated with HCl; it goes into solution without evolution of gas, even upon heating. What is the composition of the unknown?

101. Write balanced net ionic equations for the reactions, if any, that occur when:
    a. Dilute solutions of $BaCl_2$ and $NaOH$ are mixed.
    b. A strong acid is added to a precipitate of $BaCO_3$.
    c. A strong acid is added to a precipitate of $BaHPO_4$.
    d. Aqueous ammonia is added to a solution containing Ca^{2+} ions.
    e. Aqueous ammonia is added to a solution containing Mg^{2+} ions.
    f. A water solution of ammonia is acidified.

102. Describe briefly how you would distinguish between the following:
    a. Ba^{2+} and Mg^{2+} f. $BaCO_3(s)$ and $BaSO_4(s)$
    b. Ca^{2+} and NH_4^+ g. $BaCrO_4(s)$ and $BaSO_4(s)$
    c. Ba^{2+} and Na^+ h. $NH_4Cl(s)$ and $NaCl(s)$
    d. Mg^{2+} and K^+ i. NH_3 and NH_4^+
    e. K^+ and NH_4^+ j. KNO_3 and KCl

103. Describe briefly how you would distinguish between the following:
    a. Na^+ and Ag^+ e. $NH_4Cl(s)$ and $AgCl(s)$
    b. Ba^{2+} and Sb^{3+} f. $KNO_3(s)$ and $Bi(NO_3)_3(s)$
    c. Mg^{2+} and Co^{2+} g. $Ca(OH)_2(s)$ and $CaO(s)$
    d. S^{2-} and SO_4^{2-} h. $BaCrO_4(s)$ and $PbCrO_4(s)$

104. Indicate how you would accomplish the following conversions:
    a. $BaCO_3(s) \longrightarrow BaCl_2(s)$ d. $Na_2CrO_4(s) \longrightarrow NaNO_3(s)$
    b. $NH_3(g) \longrightarrow NH_4Cl(s)$ e. $Mg(OH)_2(s) \longrightarrow MgCl_2(s)$
    c. $BaSO_4(s) \longrightarrow BaCO_3(s)$ f. $Na_2S(s) \longrightarrow H_2S(g)$

105. List two different reagents that bring the cation in each of the following solids
    into solution:
    a. $BaCO_3$ b. $Mg(OH)_2$ c. $Cu(OH)_2$ d. $Zn(OH)_2$ e. Sb_2S_3

106. How would you separate the following:
    a. $CaCl_2$ from CaC_2O_4?   c. $Mg(OH)_2$ from $Cu(OH)_2$?
    b. $Mg(OH)_2$ from $Ba(OH)_2$?   d. CuS from SnS?

107. Explain why:
    a. In working with a general unknown, it is necessary to use part of the original
       sample to test for NH_4^+ and Na^+.
    b. One should not boil the solution during Step 2 of the procedure.
    c. Acetic acid rather than hydrochloric acid is used in Step 3.
    d. Addition of NH_3 in Step 4 changes the color from orange to yellow.
    e. Sodium ion is frequently reported even though it is not present in a Group
       IV unknown.

108. Describe precisely what would happen in each step of the analysis of a Group
    IV unknown containing only the following:
    a. Ba^{2+}, Mg^{2+}, and K^+ b. Mg^{2+} and NH_4^+

109. Work out a simplified scheme of analysis, omitting any extra steps, for a Group
    IV unknown containing only the following:
    a. Ca^{2+} and Mg^{2+} b. Ba^{2+}, Na^+, and NH_4^+ c. K^+ and NH_4^+

110. Describe an efficient system of analysis for an unknown that contains no cations
    other than Pb^{2+}, Cu^{2+}, Fe^{3+}, Zn^{2+}, and Ba^{2+}

111. On the basis of the following observations, indicate whether each of the ions in Group IV is present, absent, or questionable.

The unknown, when treated with NH_3 and $(NH_4)_2CO_3$ and heated, gives a precipitate. The precipitate is dissolved in acetic acid; on addition of K_2CrO_4 to the solution, it turns orange but fails to give a precipitate. Treatment of the solution with NH_3 and $K_2C_2O_4$ gives a white precipitate.

Another portion of the unknown gives a violet flame in the flame test with no trace of yellow.

112. A student analyzes an unknown that he is told may contain the cations Bi^{3+}, Sn^{2+}, Cr^{3+}, Ca^{2+}, and NH_4^+. Treatment with H_2S in acidic solution gives a precipitate that is completely insoluble in NaOH. When the H_2S solution is made basic with NH_3, no precipitate forms. The solution is evaporated to dryness, heated, and the residue dissolved in HCl. The acidic solution is made basic with NH_3, treated with $(NH_4)_2CO_3$, and heated; a white precipitate forms. The solution remaining is made strongly basic with NaOH and heated; a gas is given off, which turns red litmus blue.

Which ions are present? Which are absent? Which are questionable?

113. A general unknown, which may contain any of the ions in Groups I, II, III, and IV, is analyzed with the following results.

A sample of the unknown is made strongly basic with NaOH. A pure white precipitate forms at first; this dissolves as more NaOH is added. When the NaOH solution is heated, a gas with basic properties is evolved. The solution remaining gives a strong yellow flame in the flame test; when observed through cobalt glass, the color of the flame is purplish red.

Another sample of the original unknown gives no precipitate upon addition of HCl; however, when the HCl solution is saturated with H_2S, a precipitate forms. This precipitate dissolves completely in NaOH; treatment of the resultant solution with HCl gives a precipitate that dissolves on heating with concentrated HCl. The HCl solution is boiled and treated first with aluminum wire and then with $HgCl_2$; a white precipitate forms.

A third sample of the original unknown is treated with H_2SO_4 to give a white precipitate.

Which ions are present? Which are absent? Which are questionable?

114. Devise a scheme of analysis for a solid that may contain $PbCl_2$, $Fe(OH)_3$, $Al(OH)_3$, $BaCl_2$, and $Mg(OH)_2$.

LABORATORY ASSIGNMENTS

1. Using a known sample containing each of the Group IV cations, analyze the solution by the procedure given in the text. Compare your observations with those described. Obtain a Group IV unknown and analyze it by the procedure. Report your observations and conclusions on a Group IV flow chart.

2. Your instructor will assign to you in advance an unknown that may contain only the cations indicated in one of the following sets. Develop a procedure by which you can analyze the unknown for the ions in your set. Do not include any unnecessary steps, and try to make your scheme for separation as simple as possible.

You should not restrict yourself to the standard procedures, but you should make use, where feasible, of any of the chemical properties of the cations that facilitate their resolution or identification. Draw a complete flow chart for your scheme. Test your scheme with a sample containing each of your assigned ions, and then, when you are sure that your approach works, use it to analyze an unknown that your instructor will furnish. Record your observations and conclusions on your flow chart and turn it in to your instructor for evaluation. Your set of cations will be one of the following:

    a. Mg^{2+}, NH_4^+, Fe^{3+}, Mn^{2+}, and Pb^{2+}
    b. Ca^{2+}, Na^+, Cd^{2+}, Hg^{2+}, and Ag^+
    c. Ba^{2+}, K^+, Al^{3+}, Bi^{3+}, and Pb^{2+}
    d. Ca^{2+}, Ba^{2+}, Cr^{3+}, Ni^{2+}, and Cu^{2+}
    e. Mg^{2+}, K^+, Co^{2+}, Sn^{4+}, and Hg^{2+}
    f. Ba^{2+}, Na^+, Zn^{2+}, Fe^{3+}, and Sn^{4+}
    g. Ca^{2+}, NH_4^+, Al^{3+}, Fe^{3+}, and Co^{2+}
    h. Mg^{2+}, Ca^{2+}, Cr^{3+}, Bi^{3+}, and Cu^{2+}
    i. Ba^{2+}, NH_4^+, Ni^{2+}, Mn^{2+}, and Sb^{3+}
    j. Mg^{2+}, Ba^{2+}, Cd^{2+}, Cu^{2+}, and Al^{3+}

3. You will be given an unknown that contains only one cation, chosen from those we have studied in Groups I to IV. You will be given no more than 15 minutes in the laboratory to find out which cation is present. You will receive additional credit if you identify the cation in less than 10 minutes. Before coming to laboratory, develop the procedure you will use for making your tests. Keep your approach as simple as possible, but be sure that it works. If time permits, your instructor may issue you one or more other unknowns containing a single cation.

CHAPTER 4

THE ANIONS

A BRIEF INTRODUCTION

In this chapter the qualitative analysis of the following anions will be considered:

Sulfate Group: SO_4^{2-}, CO_3^{2-}, PO_4^{3-}, CrO_4^{2-}, $C_2O_4^{2-}$, SO_3^{2-}
Chloride Group: Cl^-, Br^-, I^-, SCN^-, S^{2-}
Nitrate Group: NO_3^-, NO_2^-, ClO_3^-, $C_2H_3O_2^-$

Although we classify the anions into groups as just indicated, their qualitative analysis is not carried out by separating groups of anions from the general mixture. The analysis of anions differs from that of cations in that tests for anions are often "spot tests" rather than ones that can be performed only at some fixed point in a rather lengthy procedure. For this reason, the identification of a specific anion can usually be accomplished more quickly and conveniently than can similar identification of a particular cation in a mixture. As you might expect, however, anion tests have the limitation that in a complex mixture it may be essentially impossible, because of interferences, to establish the presence or absence of some rather common anions.

Anion analysis, like that of the cations, is performed in a series of steps. One first carries out some preliminary tests, which allow one to establish whether the solution containing the anions has the following properties:

1. Evolves a gas when acidified with a strong acid.
2. Exhibits oxidizing capacity under acidic conditions.
3. Acts as a reducing agent under acidic conditions.
4. Produces a precipitate when treated with $BaCl_2$ solution under slightly basic conditions (Sulfate Group test).
5. Forms a precipitate when treated with $AgNO_3$ solution under acidic conditions (Chloride Group test).

On the basis of the preliminary tests, which may be quite revealing because a given anion may respond to more than one test, one proceeds to definitive identification of the anions present by applying specific tests for each anion believed to be present.

The Procedures for Analysis of the anions in a given group are included in a section allotted to that group. This section also contains a brief description of the properties of the anions in the group. The reactions that are used in the identification of that anion are discussed as are other properties that might be useful for analytical purposes. In the case of the Chloride Group, which lends itself to systematic analysis, a step-by-step procedure is presented, with a flow chart summarizing the analysis scheme.

SECTION 1

PRELIMINARY TESTS FOR ANALYSIS OF ANIONS
CO_3^{2-}, SO_4^{2-}, CrO_4^{2-}, PO_4^{3-}, $C_2O_4^{2-}$, SO_3^{2-}, Cl^-, Br^-, SCN^-, S^{2-}, NO_3^-, NO_2^-, ClO_3^-, $C_2H_3O_2^-$

The scheme of anion analysis is based on a sample solution containing the equivalent of 1 ml of a 0.1M sodium salt solution of one or more of the anions just mentioned.

Preliminary Test 1. Detection of Anions of Volatile Acids. To 2 ml of sample solution add 1 ml of 6M H_2SO_4 and mix well. Note any effervescence or other changes and cautiously note the odor of any evolved vapors. The following anions form volatile acids with the indicated properties:

CO_3^{2-}	Carbon dioxide is produced, with effervescence if the solution is moderately concentrated. There is no odor. In dilute solutions there are no gas bubbles, because CO_2 is rather soluble in water.
SO_3^{2-}	Sulfur dioxide, with the odor of burning sulfur, is formed. This gas is detectable by its odor above dilute solutions.
S^{2-}	Hydrogen sulfide is produced. The odor is that of rotten eggs and it is conclusive evidence of the presence of sulfides. The odor is detectable with very dilute solutions.
$C_2H_3O_2^-$	Acetic acid is evolved, with the odor of vinegar. Dilute acetate solutions have the odor if SO_2 or H_2S is not also there. In the presence of SO_2 or H_2S, detection is difficult.

If you are unable to reach conclusions with the cold solution, warm the tube for a minute in the boiling water bath, and recheck it for odors. In a concentrated solution or a solid, all of the earlier mentioned anions except acetate show effervescence. In addition, nitrites evolve brown NO_2 gas and form a pale blue solution; chlorates give off ClO_2, a greenish-yellow gas, and form a yellow solution.

Preliminary Test 2. Detection of Oxidizing Anions.   To 1 ml of sample solution add 1 ml of 6M HCl and 1 ml of 0.1M KI. Mix well and look for the brown or yellow color of I_2, which is produced if oxidizing anions are present. If no color appears within a minute, put the test tube in the boiling water bath for a maximum of five minutes. The following anions oxidize iodide ion, with reaction rates as indicated:

CrO_4^{2-} Even in the cold, I_2 is formed instantly.
NO_2^- There is instant formation of I_2 in the cold.
ClO_3^- Reaction is slow in the cold. It proceeds moderately rapidly, within 15 seconds, in a boiling water bath.
NO_3^- There is no reaction in the cold. A noticeable yellow color appears after a minute or two in a boiling water bath.

The presence of I_2 in solution is easily established if 1 ml of CCl_4 is added to the iodine-containing solution and the mixture is shaken. Iodine is preferentially soluble in the (lower) CCl_4 layer, where it has a characteristic purple color.

Preliminary Test 3. Detection of Reducing Anions.   To 1 ml of sample solution add 0.5 ml of 6M H_2SO_4. Add 2 drops of 0.02M $KMnO_4$ and stir. If a reducing anion is present, it tends to bleach the color of the MnO_4^- ion by reducing Mn(VII) to Mn(II). In most cases the bleaching occurs very quickly. This is the case with SO_3^{2-}, SCN^-, and NO_2^-.

With the following reducing anions, the reaction is rapid, and in addition there is a change in the appearance of the solution:

I^- or Br^- The color of the solution changes to yellow.
S^{2-} A white precipitate of sulfur forms. The precipitate tends to be very finely divided and very faint.

If $C_2O_4^{2-}$ ion is present, it does not bleach permanganate in the cold, but it decolorizes the solution readily if the test tube is heated in the boiling water bath. Traces of reductants decolorize $KMnO_4$ on prolonged heating; consequently, if you suspect oxalate is present because of bleaching at the higher temperature, add 2 drops of 0.02M $KMnO_4$ to the hot bleached solution; if oxalate is there, it bleaches the solution very quickly; if not present, the bleaching occurs very slowly, if at all.

Preliminary Test 4. Detection of Anions in the Sulfate Group. To 1 ml of sample solution in a test tube add 0.5 ml of 6M HCl. Add 6M NH_3 drop by drop until the solution is basic, and then add 0.5 ml in excess. Then add 0.5 ml of 1M $BaCl_2$ and 0.5 ml of 1M $CaCl_2$. If a precipitate forms, one or more of the following anions is present: CO_3^{2-}, SO_4^{2-}, CrO_4^{2-}, PO_4^{3-}, $C_2O_4^{2-}$, or SO_3^{2-}.

Preliminary Test 5. Detection of Anions in the Chloride Group. To 1 ml of sample solution in a 30 ml beaker add 0.5 ml of 6M H_2SO_4 and 2 ml of water. (If S^{2-} ion is present, as indicated by previous tests, bring the solution to a boil and boil gently for two minutes. Add 1 ml of water and transfer the solution to a test tube. Centrifuge out any solid precipitate, and decant the liquid into a test tube.) Add 2 or 3 drops of 0.1M $AgNO_3$ to the sample. Formation of a white or yellow precipitate indicates the presence of one or more of the following anions: Cl^-, Br^-, I^-, or SCN^-. If the solution is not boiled before adding the $AgNO_3$ and sulfide is present, a black precipitate of Ag_2S forms.

Of the anions we are considering, four will not precipitate in either the Sulfate or the Chloride Group. These ions are NO_3^-, NO_2^-, ClO_3^-, and $C_2H_3O_2^-$, and form the Nitrate Group. They are identified through the Preliminary Tests 1, 2, and 3, and by procedures given for analysis of the Nitrate Group.

   Several of the anions give reactions in more than one preliminary test. This can often be useful in analysis and makes it possible to eliminate ions from consideration, particularly when only a few anions are present. Each of the anions to be considered and the preliminary tests that would be expected to give indication of its presence are as follows:

CO_3^{2-}	1,4	SO_3^{2-}	1,3,4	S^{2-}	1,3,5
SO_4^{2-}	4	Cl^-	5	NO_3^-	2
CrO_4^{2-}	2,4	Br^-	3,5	NO_2^-	(1),2,3
PO_4^{3-}	4	I^-	3,5	ClO_3^-	(1),2
$C_2O_4^{2-}$	3,4	SCN^-	3,5	$C_2H_3O_2^-$	1

It is also useful to recognize that some anions cannot coexist stably with others. In general, *the oxidizing anions* (from Test 2) *are not present with reducing anions* (from Test 3); nitrates and oxalates may violate this rule. Sulfites and nitrites are not very stable in solution; sulfites tend to be contaminated by sulfate, formed by air oxidation, and nitrites often contain nitrate.

   Your sample solutions will be freshly prepared, but you should be aware of some of these possibilities. After completing the preliminary tests you may be able to say definitely that some anions are present or absent. More likely, you will know that one or more anions in given groups are present, but you will not know which ones. To resolve this question, proceed to the appropriate identification tests for the anions given in the Procedure for Analysis for each anion group. For the most part, there are spot tests for each anion, but in the case of analysis of the Chloride Group a systematic procedure for the whole group is employed.

COMMENTS ON PRELIMINARY TESTS
FOR ANION ANALYSIS

Test 1. Of the 15 anions studied in this scheme of analysis, 6 react to form volatile gases in acid solution. Of all these gases, only CO_2, produced from carbonate ion, has no identifying odor. The gases SO_2, H_2S, $HC_2H_3O_2$, NO_2, and ClO_2 all have characteristic smells, and they can be easily identified if no other gases are present. If two or more gases are present, the identification is more difficult. With dilute solutions, about 0.1M, there is essentially no effervescence, but the odors of some of the gases are still easily detected. In general, in cold dilute solution, nitrites and chlorates do not give noticeable reactions with acid; on heating there may be evidence of evolution of NO_2 or ClO_2.

Test 2. Oxidation of iodide in acid solution is characteristic of the oxidizing anions and has the advantage over other tests that are sometimes used in that the rate of reaction gives some indication of which oxy-anion is present. The iodide oxidizes very slowly in the hot solution even in the absence of oxidizing anions. If nitrate is indicated, it is best also to conduct the test on a pure water sample and to compare the colors of that sample and the unknown.

Test 3. Nitrite ion is the only oxy-anion we are studying with both oxidizing and reducing capabilities. If nitrite is present, it is likely that, except for nitrate, no other oxidizing or reducing anions are present. Of all the reducing anions, only oxalate requires hot acid solution for reduction of permanganate ion. If one adds $KMnO_4$ in slight excess to a cold mixture of reducing anions, all except oxalate ion react. Then if bleaching occurs on heating the solution, oxalate ion is probably also present. Chlorides react very slowly, if at all, with MnO_4^- ion under the conditions of this test.

Test 4. The advantage of the anion Group Tests is that, if a test is negative, a large number of anions can be eliminated from consideration. A positive test result is useful, but not so informative.

Test 5. If sulfide is absent from the solution, it is not necessary to boil it before adding the $AgNO_3$. Boiling readily removes S^{2-} as H_2S. Do not boil the solution to near dryness, because all the hydrogen halides are somewhat volatile, especially in boiling solutions of concentrated H_2SO_4.

PROBLEMS

1. Write balanced net ionic equations to explain the following observations:
   a. A gas is evolved when a solution containing CO_3^{2-} is acidified.
   b. A gas with a foul odor is produced when a solution of NaHS is acidified.
   c. A white precipitate forms when a solution containing SO_4^{2-} is treated with $BaCl_2$.
   d. A white precipitate forms when solutions of $Na_2C_2O_4$ and $CaCl_2$ are mixed.
   e. A yellow precipitate forms when $AgNO_3$ is added to a solution of CaI_2.

2. Write balanced net ionic equations for the following oxidation-reduction reactions that occur in acidic solution:
   a. A solution of KI is treated with hot HNO_3 (assuming NO_3^- is reduced to NO).
   b. Potassium iodide is added to an unknown containing NO_2^- (assuming reduction to NO).
   c. Potassium permanganate is added to a solution of KI (assuming reduction to Mn^{2+}).
   d. Potassium permanganate is added to a solution of Na_2S; a yellowish white solid forms.
   e. Potassium permanganate oxidizes $C_2O_4^{2-}$ to CO_2.

3. Explain briefly how you would distinguish between the following:
   a. SO_3^{2-} and Cl^-
   b. CO_3^{2-} and $C_2H_3O_2^-$
   c. CrO_4^{2-} and ClO_3^-
   d. NO_3^- and Br^-
   e. SCN^- and NO_3^-
   f. SO_4^{2-} and CrO_4^{2-}
   g. I^- and Cl^-
   h. Cl^- and SO_4^{2-}
   i. SO_3^{2-} and SO_4^{2-}
   j. $C_2H_3O_2^-$ and PO_4^{3-}

4. Indicate how you would accomplish the following conversions:
   a. $I^- \longrightarrow I_2$
   b. $CO_3^{2-} \longrightarrow CO_2(g)$
   c. $SO_3^{2-} \longrightarrow SO_4^{2-}$
   d. $NO_3^- \longrightarrow NO(g)$

5. Which of the following pairs of ions would *not* be found together in a solution?
   a. MnO_4^- and I^-
   b. I^- and Br^-
   c. ClO_3^- and NO_2^-
   d. NO_2^- and NO_3^-
   e. SO_4^{2-} and CO_3^{2-}
   f. SCN^- and NO_3^-
   g. I^- and Fe^{3+}

6. Explain the following:
   a. A negative result in Test 4 or 5 is more helpful than a positive result.
   b. The $C_2H_3O_2^-$ ion may not be detected in Test 1, even if present.
   c. Carbon tetrachloride is used in Test 2.
   d. Ammonia is used in Test 4.

7. Describe exactly what would be observed in each preliminary test with unknowns containing only the following anions:
   a. CO_3^{2-}, $C_2O_4^{2-}$, and I^- b. SO_4^{2-}, Cl^-, and NO_3^-

8. What could you deduce from the following results of preliminary tests carried out on general anion unknowns:
   a. A gas with a sharp, pungent odor is evolved in Test 1.
   b. In Test 2, I_2 is produced only on heating.
   c. In Test 3, $KMnO_4$ is not decolorized.
   d. In Test 5, a yellow precipitate forms.

9. An unknown can contain only the anions CO_3^{2-}, S^{2-}, Cl^-, and SCN^-. On addition of acid to the unknown, an odorless gas is evolved. Addition of $AgNO_3$ to the unknown causes a white precipitate to form.

   Which ions are present? Which are absent? Which are questionable? How could you check for the presence of the ions listed as questionable?

10. Using only preliminary tests, devise a scheme of analysis of an unknown containing only the following anions: SO_4^-, ClO_3^-, I^-, and S^{2-}.

101. Write balanced net ionic equations for the reactions that occur when:
   a. A solution of Na_2SO_3 is acidified.
   b. A solution of $NaC_2H_3O_2$ is acidified.
   c. An unknown containing CrO_4^{2-} is treated with $BaCl_2$.
   d. Solutions of $BaCl_2$ and Na_2SO_3 are mixed.
   e. Silver nitrate is added to an unknown containing SCN^-.

102. Complete and balance the following oxidation-reduction equations:
   a. $I^- + CrO_4^{2-} + H^+ \longrightarrow I_2 + Cr^{3+} + H_2O$
   b. $I^- + ClO_3^- + H^+ \longrightarrow I_2 + Cl^-$
   c. $MnO_4^- + Br^- \longrightarrow Mn^{2+} + Br_2$ (acidic solution)
   d. $MnO_4^- + I^- \longrightarrow$ (acidic solution)
   e. $MnO_4^- + NO_2^- \longrightarrow Mn^{2+} + NO_3^-$ (acidic solution)

103. Explain briefly how you would distinguish between the following:
   a. CO_3^{2-} and S^{2-} f. $C_2O_4^{2-}$ and SO_4^{2-}
   b. SO_3^{2-} and I^- g. Cl^- and S^{2-}
   c. NO_3^- and NO_2^- h. SCN^- and SO_4^{2-}
   d. ClO_3^- and Cl^- i. SO_3^{2-} and SO_4^{2-}
   e. NO_3^- and I^- j. I^- and Cl^-

104. How would you accomplish the following conversions?
   a. $NO_2^- \longrightarrow NO_3^-$ c. $CrO_4^{2-} \longrightarrow Cr^{3+}$
   b. $S^{2-} \longrightarrow H_2S(g)$ d. $S^{2-} \longrightarrow S(s)$

105. Which of the following pairs of ions would you not expect to find in the same solution?
   a. Br^- and Cl^- e. MnO_4^- and NO_3^-
   b. Ba^{2+} and $C_2O_4^{2-}$ f. MnO_4^- and NO_2^-
   c. Ca^{2+} and SO_4^{2-} g. I^- and NO_2^-
   d. I^- and ClO_3^-

106. Explain why:
   a. The MnO_4^- ion rather than some other strong oxidizing agent is used in Test 3.
   b. The I^- ion rather than some other easily oxidizable species is used in Test 2.
   c. A precipitate might be formed when HCl is added in Test 4.
   d. The solution is boiled before carrying out Test 5.

107. Describe precisely what would be observed in each preliminary test with unknowns containing only the following anions:
   a. CrO_4^{2-}, Cl^-, and $C_2H_3O_2^-$ b. SCN^-, S^{2-}, and Br^-

108. What can you deduce from the following results of preliminary tests carried out on general anion unknowns?
   a. No gas bubbles are observed in Test 1.
   b. In Test 2, I_2 is formed instantly in the cold.
   c. In Test 3, the purple color of MnO_4^- fades and a precipitate forms.
   d. In Test 4, a yellow precipitate forms.

109. An unknown can contain no anions other than the $C_2H_3O_2^-$, NO_3^-, SO_4^{2-}, and I^- anions. When the unknown is received, it is found to be strongly acidic and odorless. When KI is added to a sample of the unknown, oxidation is observed upon heating. Addition of $BaCl_2$ to a sample of the unknown gives a white precipitate.
   Which ions are present? Which are absent?

110. Using only preliminary tests, describe how you would analyze an unknown containing no anions other than the CO_3^{2-}, $C_2H_3O_2^-$, CrO_4^{2-}, NO_3^-, and Cl^- anions.

LABORATORY ASSIGNMENTS

1. You will be given an unknown solution containing only one anion. Carry out the five preliminary tests on the solution, recording your observations and conclusions. If possible, on the basis of your observations, determine which anion is present. If this is not possible, report which anions may be present. Your instructor may then give you another solution to analyze.

2. Carry out the preliminary tests on an unknown solution that contains two anions, each of which can be unambiguously identified in the presence of the other by its responses to these tests. Report your observations and conclusions to your instructor.

SECTION 2

THE PROPERTIES OF THE ANIONS IN THE SULFATE GROUP—CO_3^{2-}, SO_4^{2-}, CrO_4^{2-}, PO_4^{3-}, $C_2O_4^{2-}$, SO_3^{2-}

CARBONATE, CO_3^{2-}

The acid associated with the carbonate ion is carbonic acid, H_2CO_3, made by dissolving carbon dioxide in water. Carbonic acid is a weak acid, which must dissociate twice to produce the carbonate ion:

$$CO_2 + H_2O = (H_2CO_3) \rightleftharpoons H^+ + HCO_3^- \qquad K_1 = 4.2 \times 10^{-7}$$
$$HCO_3^- \rightleftharpoons H^+ + CO_3^{2-} \qquad K_2 = 4.8 \times 10^{-11}$$

There is some question about whether H_2CO_3 molecules actually exist in the water solution; pure H_2CO_3 cannot be recovered, and it is perhaps just as well to consider that if one sees the symbol H_2CO_3, it means that we have a solution of CO_2 dissolved in water and that the concentration of the CO_2 is equal to the concentration of the carbonic acid. Carbon dioxide is moderately soluble in water, about 0.035 moles/lit at 25° C under 1 atm pressure of CO_2.

Many carbonate salts are insoluble in water. Because they are salts of a weak acid, we can expect carbonates to be soluble in acidic solution. This is indeed the case, with the evolution of CO_2 being a prime characteristic reaction of carbonates on treatment with strong acids.

Monohydrogen carbonate salts of the alkali metals are sometimes encountered, with $NaHCO_3$ being the common example. Solutions of the salt in water are slightly basic. Sodium carbonate can be made by heating $NaHCO_3$ to about 250° C:

$$2\,NaHCO_3(s) \longrightarrow Na_2CO_3(s) + H_2O(g) + CO_2(g)$$

CHARACTERISTIC REACTIONS OF THE CARBONATE ION

1. **Strong Acids.** Carbonates in general effervesce when treated with acidic solutions, and they evolve CO_2 as CO_3^{2-} ion decomposes. The reaction occurs both with moderately concentrated solutions and with solid salts:

$$CO_3^{2-} + 2H^+ \longrightarrow CO_2(g) + H_2O$$
$$CaCO_3(s) + 2\,H^+ \longrightarrow Ca^{2+} + CO_2(g) + H_2O$$

The confirmatory test for the carbonate ion is to pass the CO_2 given off in the reaction just mentioned into a solution of barium hydroxide; the solution becomes cloudy as the carbonate is formed:

$$CO_2(g) + Ba^{2+} + 2\,OH^- \longrightarrow BaCO_3(s) + H_2O$$

2. **Barium Chloride.** Solutions containing carbonate ion yield a white precipitate of $BaCO_3$ on being treated with solutions containing Ba^{2+} ion. The precipitate is soluble in acid solutions.

$$Ba^{2+} + CO_3{}^{2-} \rightleftharpoons BaCO_3(s)$$

3. **Silver Nitrate.** In the presence of $CO_3{}^{2-}$, this reagent produces a white precipitate of Ag_2CO_3, which on boiling becomes dark because of formation of Ag_2O:

$$Ag_2CO_3(s) \longrightarrow Ag_2O(s) + CO_2(g)$$

SULFATE, $SO_4{}^{2-}$

Sulfate salts are typically soluble in water and commonly available in chemical stockrooms. Among the common sulfates, only $PbSO_4$ and $BaSO_4$ are insoluble, and, as might be expected, these two salts are often formed in sulfate analysis.

Sulfate salts may be considered to be derived from the strong acid, H_2SO_4. Sulfuric acid can be formed by dissolving SO_3 in water. However, unlike carbonic acid, a solution of sulfuric acid, even when concentrated, exhibits no appreciable vapor pressure of the dissolved oxide. To drive SO_3 from sulfuric acid one must boil off most of the water and then, finally, at temperatures exceeding 300° C, SO_3 begins to evolve as a choking gas.

Sulfuric acid is a strong acid. Its first ionization, to form $HSO_4{}^-$ and H^+ is essentially complete; its second ionization, to produce sulfate ion, has a dissociation constant of about 1×10^{-2}, making $HSO_4{}^-$ a moderately strong acid.

CHARACTERISTIC REACTIONS OF THE SULFATE ION

1. **Barium Chloride.** Solutions containing sulfate ion yield a white precipitate of barium sulfate on treatment with this reagent. Barium sulfate is not soluble in water or dilute acids. Since $BaSO_4$ is the only compound of barium with this property, it is used in the classic test for the presence of sulfate ion.

2. **Lead Nitrate.** White $PbSO_4$ precipitates when solutions containing Pb^{2+} and SO_4^{2-} ions are mixed. Lead sulfate is very insoluble in water, but it can be brought into solution in strong NaOH or by hot concentrated ammonium acetate, $NH_4C_2H_3O_2$.

CHROMATE, $CrO_4{}^{2-}$

Chromate salts are all colored, and they are usually yellow or red. Most of them are insoluble in water but soluble in 6M HNO_3. The laboratory source of chromate ion is usually K_2CrO_4 or Na_2CrO_4, both of which are water soluble, yellow, crystalline materials. Solutions of chromates are characteristically yellow, turning orange when acidified, because of formation of $Cr_2O_7^{2-}$, the dichromate ion:

$$2\,CrO_4{}^{2-} + 2\,H^+ \rightleftharpoons Cr_2O_7{}^{2-} + H_2O$$

Chromate salts might be considered to be derivatives of $H_2Cr_2O_7$. A solution of $H_2Cr_2O_7$ can be prepared by dissolving CrO_3 in water, but the pure acid cannot be recovered. Of course, H_2CrO_4 cannot be prepared because of the reaction of $CrO_4{}^{2-}$ ion with acids.

CHARACTERISTIC REACTIONS OF THE CHROMATE ION

1. **Barium Chloride.** Adding a solution containing Ba^{2+} to a chromate produces a yellow precipitate of $BaCrO_4$, which is soluble in strong acids:

$$2\,BaCrO_4(s) + 2\,H^+ \rightleftharpoons 2\,Ba^{2+} + Cr_2O_7{}^{2-} + H_2O$$

2. **Hydrogen Peroxide.** In the presence of acid, a 3 per cent solution of H_2O_2 reacts with chromates to form a blue chromium peroxide, thought to have the formula CrO_5:

$$Cr_2O_7{}^{2-} + 4\,H_2O_2 + 2\,H^+ \longrightarrow 2\,CrO_5 + 5\,H_2O$$

The blue peroxide is unstable and decomposes to Cr^{3+}. If one adds a little ether before adding the H_2O_2 to the acidified chromate, and then shakes the solution gently when the blue color forms, the peroxide is extracted into the ether and the color persists for a much longer time.

3. **Lead Nitrate.** Addition of this reagent to solutions of $CrO_4{}^{2-}$ yields a very insoluble yellow precipitate of $PbCrO_4$, which can be brought into solution in 6M NaOH:

$$PbCrO_4(s) + 4\,OH^- \rightleftharpoons Pb(OH)_4{}^{2-} + CrO_4{}^{2-}$$

4. **Silver Nitrate.** When $AgNO_3$ is mixed with CrO_4^{2-}, a red precipitate of Ag_2CrO_4 is formed, which is soluble in 6M NH_3 and in 6M HNO_3.

5. **Potassium Iodide.** In acid solution, iodide ion reduces dichromate ion, liberating brown or orange iodine into the system:

$$Cr_2O_7^{2-} + 6\,I^- + 14\,H^+ \longrightarrow 2\,Cr^{3+} + 3\,I_2 + 7\,H_2O$$

PHOSPHATE, $PO_4{}^{3-}$

Phosphate salts may be considered to be derived from phosphoric acid, H_3PO_4. Because H_3PO_4 is a weak acid, ionizing in steps $(K_1 = 1 \times 10^{-2}$, $K_2 = 2 \times 10^{-7}$, $K_3 = 3.5 \times 10^{-13})$, it is not surprising that one finds salts with the formulas Na_3PO_4, Na_2HPO_4, and NaH_2PO_4, all available at the stock room. The formula of a sodium phosphate salt depends on the pH of the solution from which it is crystallized. In strongly acid solutions, the salt obtained would be NaH_2PO_4; in a neutral solution, Na_2HPO_4 would be the product, whereas Na_3PO_4 crystallizes from strongly basic solution.

Phosphoric acid may be prepared by dissolving solid P_4O_{10} in water. The pure acid is a deliquescent solid. Phosphoric acid solutions do not give off any phosphorus oxide vapors on being heated because phosphoric acid is not volatile. It also shows little tendency to undergo oxidation-reduction reactions.

CHARACTERISTIC REACTIONS OF THE PHOSPHATE ION

1. **Barium Chloride.** Depending on the pH, white insoluble $BaHPO_4$ or $Ba_3(PO_4)_2$ precipitates when barium and phosphate solutions are mixed. The precipitate is soluble in dilute acids.

2. **Silver Nitrate.** Yellow silver phosphate, Ag_3PO_4, precipitates on addition of this reagent to PO_4^{3-} ion. The substance is soluble in 6M HNO_3 and in 6M NH_3.

3. **Lead Nitrate.** A white lead phosphate, $Pb_3(PO_4)_2$, precipitates when this reagent is added to PO_4^{3-} ion. This salt goes into solution in 6M HNO_3.

4. **Ammonium Molybdate.** In hot nitric acid solution phosphates react with ammonium molybdate to form a characteristic yellow precipitate of ammonium phosphomolybdate, $(NH_4)_3PO_4 \cdot 12\,MoO_3$:

$$H_2PO_4^- + 12\,MoO_4^{2-} + 3\,NH_4^+ + 22\,H^+ \longrightarrow$$
$$(NH_4)_3PO_4 \cdot 12\,MoO_3(s) + 12\,H_2O$$

The precipitate is soluble in excess phosphoric acid and in 6M NaOH and

6M NH_3. Reducing anions may interfere with this test by reacting with the $MoO_4{}^{2-}$, producing a blue color. These anions may be oxidized with strong HNO_3 before making the test.

SULFITE, $SO_3{}^{2-}$

Sulfites are the salts of sulfurous acid, H_2SO_3, made by dissolving SO_2 gas in water. The parent acid, H_2SO_3, is moderately strong ($K_1 = 2 \times 10^{-2}$, $K_2 = 5 \times 10^{-6}$). Sulfurous acid cannot be isolated, and one usually assumes that 0.1M H_2SO_3 means 0.1 moles SO_2 dissolved per liter of solution, and not that actual H_2SO_3 molecules exist in the solution:

$$SO_2 + H_2O = (H_2SO_3) \rightleftharpoons H^+ + HSO_3{}^-$$

The situation is completely analogous to that with carbonic acid and with ammonia, in which neither H_2CO_3 nor NH_4OH can be isolated.

A solution of sulfurous acid, or an acidified sulfite, evolves SO_2, which may be easily detected by its very characteristic sharp odor. Boiling an acidified sulfite for a few minutes removes essentially all the SO_2 from the solution.

All sulfites except those of the alkali metals and $NH_4{}^+$ are insoluble in water; the latter are therefore the usual laboratory sources of sulfite ion. Monohydrogen sulfites, salts of the $HSO_3{}^-$ ion, also are known and are more soluble than the sulfites.

In addition to the often associated odor of SO_2, sulfites are characterized by being readily oxidized to sulfates. The reaction tends to occur in moist air and proceeds rapidly with many oxidizing agents.

CHARACTERISTIC REACTIONS OF THE SULFITE ION

1. **Sulfuric or Hydrochloric Acid.** Non-oxidizing acids decompose sulfite solutions or salts with the evolution of SO_2:

$$SO_3{}^{2-} + 2 H^+ \rightleftharpoons (H_2SO_3) = SO_2(g) + H_2O$$

2. **Hydrogen Peroxide.** In acid solution, H_2O_2 oxidizes sulfites to form sulfates:

$$H_2O_2 + (H_2SO_3) \longrightarrow SO_4{}^{2-} + H_2O + 2 H^+$$

3. **Barium Chloride.** A precipitate of white $BaSO_3$ forms with neutral or basic

solutions of sulfite ion. The precipitate is very easily soluble in nitric or hydrochloric acid:

$$BaSO_3(s) + 2\,H^+ \rightleftharpoons Ba^{2+} + SO_2(g) + H_2O$$

4. **Silver Nitrate.**   This reagent precipitates white Ag_2SO_3, which dissolves in ammonia, nitric acid, or excess sulfite, in the last case by formation of a sulfite complex ion.

5. **Potassium Permanganate.**   In dilute sulfuric acid solution, permanganate solution is decolorized by sulfites, with the formation of sulfate ion and Mn^{2+}. Dichromate ion, iodine, and Fe^{3+} are also reduced by sulfurous acid:

$$5\,(H_2SO_3) + 2\,MnO_4^- + H^+ \longrightarrow 5\,HSO_4^- + 2\,Mn^{2+} + 3\,H_2O$$

OXALATE, $C_2O_4{}^{2-}$

Oxalate salts are typically insoluble in water, the exceptions being those of the alkali metals, magnesium, and chromium. All oxalates dissolve in 6M HCl. In solution the oxalate ion is colorless. It can be slowly oxidized, and converted to CO_2, by hot, strong oxidizing agents.

Oxalic acid, $H_2C_2O_4$, is an organic acid ($K_1 = 4 \times 10^{-2}$ and $K_2 = 5 \times 10^{-5}$); a solution of $0.1M\ H_2C_2O_4$ is about 50 per cent ionized. The crystalline acid exists as a dihydrate and can be sublimed without decomposition if heated carefully. When strongly ignited, oxalic acid, and oxalates in general, decompose with the evolution of CO_2.

The oxalate ion is a good complexing agent for metallic cations. It is used to complex Sn^{4+} in the Group II analysis to prevent precipitation of SnS_2 in the presence of antimony. It also forms stable complexes with Fe^{3+} and Cu^{2+}.

CHARACTERISTIC REACTIONS OF THE OXALATE ION

1. **Barium or Calcium Chloride.**   A white precipitate of $BaC_2O_4 \cdot H_2O$ or $CaC_2O_4 \cdot H_2O$ forms on addition of these reagents. These salts are soluble in strong acids; the barium compound also goes into solution in acetic or oxalic acid. The most insoluble oxalate is $CaC_2O_4 \cdot H_2O$, which can be precipitated from acetic acid solution.

2. **Potassium Permanganate.**   In hot acid solution, the purple MnO_4^- ion is decolorized by oxalates, with the formation of CO_2 and Mn^{2+}:

$$5\,H_2C_2O_4 + 2\,MnO_4^- + 6\,H^+ \longrightarrow 2\,Mn^{2+} + 10\,CO_2 + 8\,H_2O$$

PROCEDURE FOR ANALYSIS OF
ANIONS IN THE SULFATE GROUP

Unless otherwise directed, you may assume that 5 ml of sample solution may contain the equivalent of 1 ml of 0.1M solutions of the sodium salt of one or more of the anions in the sulfate group.

Step 1. *Test for the presence of carbonate.* Place the sample, 2 ml of solution or 0.1 g of solid, in a regular, not semi-micro, test tube. Add 1 ml of 3 per cent H_2O_2 and 1 ml of 6M HCl and stir well. Using the apparatus illustrated in Figure 4-1, pass any gas that is given off into 3 ml of a clear, saturated solution of $Ba(OH)_2$. Heat the sample gently until no more bubbles are observed passing into the hydroxide solution. A white precipitate or cloudiness due to $BaCO_3$ shows the presence of carbonate.

The rubber stopper must fit the large test tube tightly. Hold your finger over the side arm while the gas is being evolved and during the heating. To keep the gas from backing up into the large tube, lift your finger when

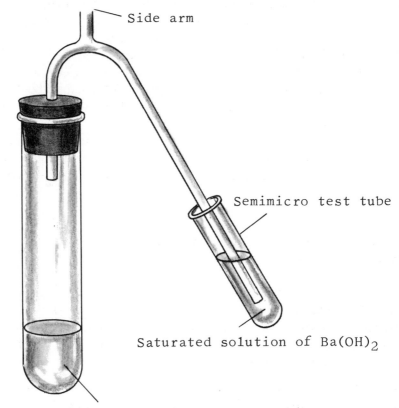

Side arm

Semimicro test tube

Saturated solution of $Ba(OH)_2$

Carbonate test solution

FIGURE 4-1. Apparatus for detection of carbonate ion.

necessary during the heating process. Carbonate is often present in basic solutions as an impurity, because such solutions tend to absorb, and react with, CO_2 from the air. If your test results are doubtful, compare your results with those you obtain with 1 ml of 0.1M Na_2CO_3 solution.

Step 2. *Test for the presence of sulfate.* Pour 1 ml of sample solution into a test tube and add 1 ml of 6M HCl. Add a few drops of 1M $BaCl_2$ and stir well. If sulfate ion is present, a white, finely divided precipitate of $BaSO_4$ forms. Traces of sulfate produce cloudiness; therefore, if the test results are doubtful, compare the amount of precipitate you obtained with that from 1 ml of 0.001M H_2SO_4, which contains about 0.1 mg of SO_4^{2-}, prepared by dilution of a stock sample of acid.

Step 3. *Test for the presence of chromate.* If the sample solution is colorless, it does not contain chromate. If it is yellow or orange, it may. To 1 ml of solution add 1 ml of 6M HNO_3. If the solution turns orange, cool it under the tap and then add 1 ml of 3 per cent H_2O_2. A blue coloration, which may fade rapidly, is proof of the presence of chromate. If the solution turns green or blue, chromate ion is present and is being reduced to Cr^{3+} by another species in the acidic solution.

Step 4. *Test for the presence of phosphate.* To 1 ml of sample solution in a test tube add 1 ml of 6M HNO_3. Add 1 ml of 0.5M $(NH_4)_2MoO_4$, and stir thoroughly. Put the test tube in a boiling water bath for a few minutes. Remove the test tube from the bath and let it stand for at least 10 minutes. A yellow precipitate of ammonium phosphomolybdate, which may form quite slowly, establishes the presence of phosphate. Reducing anions may interfere with this test; see comments.

Step 5. *Test for the presence of sulfite.* In the preliminary test with acid, sulfites evolve SO_2, which can be detected by its odor at very low concentrations. To confirm the presence of sulfite, add 1 ml of 6M HCl and 1 ml of 1M $BaCl_2$ to 1 ml of sample solution. Stir the solution well and centrifuge out any precipitate of $BaSO_4$. Decant the clear solution into a test tube and add 1 ml of 3 per cent H_2O_2. Stir the solution and let it stand for a few minutes. If sulfite is present, its oxidation to sulfate causes a new precipitate of white $BaSO_4$ to form.

Step 6. *Test for the presence of oxalate.* Add 0.5 ml of 6M acetic acid to 1 ml of sample solution in a test tube. Then add 1 ml of 1M $CaCl_2$. Stir the solution well and allow it to stand for a maximum of 10 minutes. If a white precipitate forms, it is highly indicative of the presence of oxalate. Centrifuge out any precipitate and decant the solution, which may be discarded. Wash the solid thoroughly twice with 4 ml of water. Centrifuge and discard the wash. To the solid add 1 ml of 6M H_2SO_4 and put the test tube in the boiling water bath for a minute or so, stirring occasionally to help the solid dissolve. To the hot solution add 1 or 2 drops of 0.02M $KMnO_4$. If oxalate is present it quickly bleaches the purple solution.

COMMENTS ON THE PROCEDURE FOR
ANALYSIS OF THE SULFATE GROUP

Step 1. The addition of H_2O_2 can be omitted if sulfites are absent. This reagent oxidizes sulfite to sulfate and thereby prevents evolution of SO_2 and precipitation of barium sulfite, which would indeed interfere with the test. This test for carbonate is quite sensitive and can easily be used to detect 1 mg of carbonate.

Step 2. Sulfites usually contain enough sulfate to give a positive test result here. If you get more than a trace indication, you should report it. Surprisingly enough, this simple test is definitive evidence for sulfate in the presence of all the other anions included in the scheme.

Step 3. Some texts call for the addition of 1 ml of ether and then the H_2O_2 and shaking. This extracts the blue substance, CrO_5, a chromium peroxide, into the (upper) ether layer and prolongs the duration of its color. In general, this is not a necessary additional step as long as the chromate concentration is appreciable. Reducing anions react with chromates in acidic solution, forming green $Cr(III)$ species. If the color changes from orange to green, a chromate must be present, and, in addition, a reducing anion is present also.

Step 4. If reducing anions are present, a blue solution may form here, and there may possibly be other problems. If there appears to be trouble, get a new 1 ml sample and add 1 ml of 6M HNO_3. Boil it for a minute or two in a 30 ml beaker, replenishing the liquid with 6M HNO_3 if necessary. When you are finished, you should have a clear solution, colorless if chromate is absent, and possibly a white sulfur precipitate. Centrifuge out any precipitate and decant the liquid into a test tube. This liquid should be free of any interfering anions. Add 0.5 ml of 6M HNO_3 and 0.5 ml of 0.5M $(NH_4)_2MoO_4$ and look for the yellow precipitate as before.

Step 5. The detection of sulfite is straightforward. If SO_3^{2-} is present you should not miss it.

Step 6. In dilute solution, calcium oxalate may be slow to form. Even when it does precipitate, it may be very light. However, it does not take much solid to show the reducing properties of the oxalate ion. Since other reducing anions can decolorize permanganate, it is important to wash the calcium oxalate very carefully before adding the sulfuric acid.

PROBLEMS

1. Write balanced net ionic equations to explain the following observations:
   a. A gas is evolved when hydrochloric acid is added to solid barium carbonate.
   b. When silver nitrate is added to a solution of sodium carbonate, a white precipitate forms; upon heating, the precipitate turns black (two reactions).
   c. A solution prepared by boiling $PbSO_4$ with concentrated Na_2CO_3 gives a precipitate when treated with $BaCl_2$ in acidic solution (two reactions).
   d. Lead chromate goes into solution when treated with NaOH.
   e. Silver phosphate dissolves upon the addition of either H^+ or NH_3 (two reactions).
   f. Silver sulfite dissolves in acid.

2. Write net ionic equations for the oxidation-reduction reactions that occur when:
   a. Hydrogen peroxide is added to an acidic solution containing chromium in the $+6$ oxidation state.
   b. Hydrogen peroxide is added to a solution of SO_3^{2-}.
   c. Solutions containing MnO_4^- and $HC_2O_4^-$ are mixed (acidic solution).

3. Indicate how you would distinguish between the following:
   a. CO_3^{2-} and SO_4^{2-}
   b. CrO_4^{2-} and $Cr_2O_7^{2-}$
   c. PO_4^{3-} and SO_3^{2-}
   d. $Na_2C_2O_4(s)$ and $CaC_2O_4(s)$
   e. $Ag_2SO_3(s)$ and $Ag_2S(s)$
   f. $NaNO_3(s)$ and $Na_3PO_4(s)$

4. List two reagents that dissolve the following:
   a. $PbCrO_4$ b. Ag_2CrO_4 c. Ag_2SO_3 d. BaC_2O_4

5. Explain why:
   a. Hydrogen peroxide is added in Step 1.
   b. The SO_4^{2-} ion is usually found in solutions containing SO_3^{2-}.
   c. Solutions containing I^- and CrO_4^{2-} turn green on addition of HNO_3.
   d. Calcium chloride rather than $BaCl_2$ is used in Step 6.

6. How would the results of the analysis of the sulfate group be affected by the following:
   a. Omitting HNO_3 in Step 4?
   b. Using HNO_3 instead of HCl in Step 2?
   c. Omitting HCl in Step 5?
   d. Using MnO_2 instead of $KMnO_4$ in Step 6?

7. Describe a simple scheme of analysis for an unknown containing only the following:
   a. CO_3^{2-} and $C_2O_4^{2-}$ b. SO_4^{2-} and SO_3^{2-}

8. An unknown that may contain any of the sulfate group anions but no other anions is tested with the following results.

   Addition of HCl to the solution causes evolution of a gas. Addition of NH_3 followed by $BaCl_2$ results in formation of a precipitate that is completely soluble in acid with the evolution of a gas. Treatment of the unknown with $CaCl_2$ and acetic acid produces a precipitate that, after treatment with H_2SO_4 and $KMnO_4$, gives a deeply colored solution.

   Which ions are present? Which are absent? Which are questionable?

9. Devise a scheme of analysis for a solid that may contain Ag_2CO_3, $BaSO_4$, CaC_2O_4, and Na_2SO_4.

101. Write balanced net ionic equations for the reactions that occur when:
    a. Carbon dioxide is passed into a solution of calcium hydroxide.
    b. Solutions of lead acetate and sodium sulfate are mixed.
    c. Barium chromate is treated with hydrochloric acid.
    d. Solid $BaHPO_4$ is treated with dilute HCl.
    e. Hydrochloric acid is added to a solution of Na_2SO_3.
    f. Silver nitrate is added to a solution of Na_2SO_3.
    g. Barium chloride is added to a solution of sodium oxalate.

102. Complete and balance the following oxidation-reduction equations:
    a. $Cr_2O_7^{2-} + Br^- \longrightarrow Cr^{3+} + Br_2$ (acidic solution)
    b. $MnO_4^- + SO_3^{2-} + H^+ \longrightarrow Mn^{2+} + SO_4^{2-} + H_2O$
    c. $Ag_2S_2O_3(s) + H_2O \longrightarrow Ag_2S(s) + SO_4^{2-} + H^+$

103. Describe a simple test that would enable you to distinguish between the following:
    a. CO_3^{2-} and CrO_4^{2-} d. $Na_2CO_3(s)$ and $NaHCO_3(s)$
    b. Ba^{2+} and Mg^{2+} e. $Na_2SO_3(s)$ and $Na_2SO_4(s)$
    c. SO_4^{2-} and PO_4^{3-} f. $Na_2SO_3(s)$ and $NaHSO_3(s)$

104. Give the formula of a species that would dissolve:
    a. $BaCO_3$ but not $BaSO_4$ c. $BaSO_3$ but not $BaSO_4$
    b. $PbSO_4$ but not $BaSO_4$ d. Ag_2SO_4 but not $PbSO_4$

105. Explain why:
    a. One may obtain a weak test for CO_3^{2-} with an unknown to which no CO_3^{2-} was added.
    b. Neither $BaSO_3$ nor BaC_2O_4 precipitates in Step 2.
    c. Hydrogen peroxide is added in Step 5.

106. What effect, if any, would the following substitutions in reagents have on the various tests for sulfate group anions?
    a. Sodium hydroxide for $Ba(OH)_2$ in the test for CO_3^{2-}.
    b. Water for H_2O_2 in the test for CrO_4^{2-}.
    c. Ammonia for acetic acid in Step 6.

107. Describe a simple scheme of analysis for an unknown containing no anions other than the following:
    a. CrO_4^{2-}, SO_4^{2-}, and PO_4^{3-} b. CO_3^{2-}, SO_3^{2-}, and CrO_4^{2-}

108. An unknown that may contain any of the sulfate group anions but no other anions is tested with the following results.
    On addition of HCl, a pungent gas is evolved, which gives a precipitate when passed into $Ba(OH)_2$. Addition of $BaCl_2$ to the acidic solution gives no precipitate. However, when the solution is made basic, a yellowish precipitate forms.
    Which ions are present? Which are absent? Which are questionable?

109. Devise a scheme of analysis for a solid that may contain $Na_2C_2O_4$, $CaCO_3$, $BaCrO_4$, and Na_3PO_4.

LABORATORY ASSIGNMENTS

1. Make up a known containing all of the anions in the Sulfate Group except SO_3^{2-}. Carry out the spot tests for each of the anions as given in the procedure and compare your observations with those described. On a separate sample, perform the test for sulfite ion. Obtain an unknown and analyze it to determine which Sulfate Group anions it contains. You may find that it saves time to go through Preliminary Tests 1 to 3 before doing any spot identifications. Report all of your observations and conclusions.

2. You will be given an unknown that contains only one anion, chosen from the Sulfate Group. Develop a procedure that allows you, in the shortest possible time, to find out which anion is present. Your procedure may include standard spot tests, but you will find it useful to consider shortcuts. Your instructor will issue you an unknown for analysis by your procedure. When you are sure which ion is present, report your result for evaluation and obtain another unknown. Your grade will depend on the number of samples you are able to analyze correctly in the time allowed.

3. You will be given a sample that may contain only those anions in one of the following sets. When you have been assigned a set of anions, develop a procedure, based on the Preliminary Tests and the Sulfate Group tests, by which you can determine which anions are present in a sample solution. Draw a flow chart in which you indicate all reagents to be used and the species that may be present in each step. Test your procedure with a known sample to make sure that it works, and then analyze an unknown that your instructor will furnish. Report your observations and conclusions on the flow chart. The sets to be considered are the following:

   a. CO_3^{2-}, SO_4^{2-}, SO_3^{2-}, and Cl^-
   b. SO_4^{2-}, CrO_4^{3-}, PO_4^{3-}, and SO_3^{2-}
   c. CO_3^{2-}, PO_4^{3-}, $C_2O_4^{2-}$, and Br^-
   d. SO_4^{2-}, $C_2O_4^{2-}$, $C_2H_3O_2^-$, and PO_4^{3-}
   e. SO_3^{2-}, SO_4^{2-}, SCN^-, and S^{2-}
   f. CO_3^{2-}, SCN^-, PO_4^{3-}, and $C_2O_4^{2-}$
   g. SO_3^{2-}, I^-, SO_4^{2-}, and $C_2O_4^{2-}$
   h. CO_3^{2-}, CrO_4^{2-}, SO_3^{2-}, and Cl^-

SECTION 3

THE PROPERTIES OF THE ANIONS IN THE CHLORIDE GROUP—Cl⁻, Br⁻, I⁻, SCN⁻, S²⁻

CHLORIDE, Cl⁻

Chlorides are among the most common commercially available salts, and, being typically soluble, they are often used as sources of needed metallic cations in solution. Only the chlorides of the cations in Group I plus the oxychlorides of antimony and bismuth are insoluble in water.

In principle, chlorides can be prepared by reaction of hydrochloric acid, HCl, with metallic oxides or hydroxides. Hydrochloric acid is a solution of hydrogen chloride, which is a gas, in water; because HCl is extremely soluble in water, one can use the commercial acid, which is 12M, without undue loss of HCl vapor. Hydrogen chloride is more volatile than SO_3 from sulfuric acid, and boiling a strong HCl solution down to a small volume vaporizes most of the HCl. If a mixture of HCl and H_2SO_4 is boiled, essentially all the chloride is removed before any SO_3 is observed. Hydrochloric acid is perhaps the acid most often used in the inorganic laboratory; it is strong, non-oxidizing, and reasonably easily removed by boiling.

Except for $HgCl_2$ and $CdCl_2$, metallic chlorides in solution are essentially completely ionized.

CHARACTERISTIC REACTIONS OF THE CHLORIDE ION

1. **Silver Nitrate.** Chlorides react with solutions containing Ag^+ ion to form a white, curdy precipitate of AgCl, which is insoluble in strong acids and bases. Silver chloride is soluble in 6M NH_3 and also dissolves in solutions containing other complexing ligands, particularly CN^- and $S_2O_3^{2-}$, or even Cl^- at very high concentrations:

$$AgCl(s) + 2\,NH_3 \rightleftharpoons Ag(NH_3)_2^+ + Cl^-$$
$$AgCl(s) + Cl^- \rightleftharpoons AgCl_2^-$$

In the presence of acid, silver chloride is reduced by metallic zinc:

$$2\,AgCl(s) + Zn(s) \longrightarrow 2\,Ag(s) + Zn^{2+} + 2\,Cl^-$$

2. **Sulfuric Acid.** When added to a solid chloride salt, concentrated H_2SO_4 causes HCl to be given off. This test must not be carried out unless strongly oxidizing anions are known to be absent, since an explosion may result:

$$H_2SO_4(l) + NaCl(s) \longrightarrow HCl(g) + NaHSO_4(s)$$

3. **Oxidizing Agents.** In strongly acid solution good oxidizing species such as $KMnO_4$, $NaBiO_3$, and MnO_2, but not $K_2S_2O_8$, oxidize chloride ion to chlorine:

$$2 Cl^- + MnO_2(s) + 4 H^+ \longrightarrow Cl_2(g) + Mn^{2+} + 2 H_2O$$

BROMIDE, Br⁻

Bromide salts resemble chlorides in most of their general properties. The metals that have insoluble chlorides have insoluble bromides; as a class, these bromides are less soluble than the chlorides.

Hydrobromic acid, HBr, the acid from which bromide salts derive, is a solution of gaseous hydrogen bromide in water. It is a strong, nonoxidizing acid, much like HCl in its properties. Like the bromide salts, it is reasonably easily oxidized by typical oxidizing agents, yielding Br_2.

CHARACTERISTIC REACTIONS OF THE BROMIDE ION

1. **Silver Nitrate.** With Br⁻ ion, this reagent yields a light yellow precipitate of AgBr. This salt is more insoluble than AgCl, but it dissolves, in 15M NH_3 and in $Na_2S_2O_3$ solution. It can be reduced by zinc metal in the presence of acid, freeing the Br⁻ ion and forming silver metal.

2. **Nitric Acid.** All bromides, except for the very insoluble AgBr, are oxidized by 15M HNO_3:

$$6 Br^- + 2 NO_3^- + 8 H^+ \longrightarrow 3 Br_2 + 2 NO(g) + 4 H_2O$$

At least part of the Br_2 remains in solution and can be extracted into CCl_4, where it has a yellow or brown color. Iodides and thiocyanates interfere with this test.

3. **Chlorine water.** Bromide ion in solution is oxidized to bromine by chlorine water. Again, extraction into CCl_4 helps one see the Br_2. As might be expected, iodides interfere.

$$2 Br^- + Cl_2 \longrightarrow Br_2 + 2 Cl^-$$

4. **Other Oxidizing Agents.** The agents $K_2S_2O_8$, $KMnO_4$, and MnO_2, in the presence of H_2SO_4, liberate bromine from bromide solutions:

$$S_2O_8{}^{2-} + 2\,Br^- \longrightarrow Br_2 + 2\,SO_4{}^{2-}$$

The reagents $K_2Cr_2O_7$ and KNO_2 do not readily oxidize bromides. Hot, concentrated H_2SO_4 added to a solid bromide causes both HBr and Br_2 to be evolved:

$$H_2SO_4(l) + KBr(s) \longrightarrow HBr(g) + KHSO_4(s)$$
$$2\,HBr(g) + H_2SO_4(l) \longrightarrow Br_2(g) + SO_2(g) + 2\,H_2O$$

IODIDE, I⁻

Except for the iodides of the Group I cations, HgI_2 and BiOI, all iodides are soluble in water. Iodides are similar to the other halide salts, but they are much more easily oxidized, especially in solution. After a time, KI solution turns yellow because of air oxidation of I^- ion. Iodide ion is a good complexing ligand, even reacting with I_2 to form the $I_3{}^-$ ion.

CHARACTERISTIC REACTIONS OF THE IODIDE ION

1. **Silver Nitrate.** Iodide ion reacts with $AgNO_3$, producing a yellow precipitate of AgI. This is the most insoluble of the silver halides, and it is not appreciably soluble in 15M NH_3. It can be decomposed by reaction with zinc metal in the presence of sulfuric acid. It also goes into solution in $Na_2S_2O_3$ or KCN, forming the complex ions $Ag(S_2O_3)_2{}^{3-}$ and $Ag(CN)_2{}^-$, respectively.

2. **Potassium Peroxydisulfate.** In warm acetic acid solution this reagent oxidizes iodide ion to iodine, but it does not affect bromides.

$$2\,I^- + S_2O_8{}^{2-} \longrightarrow I_2 + 2\,SO_4{}^{2-}$$

Detection of I_2 is easily accomplished by extraction into CCl_4, in which the I_2 has a reddish violet color. In water solution I_2 is brown, orange, or yellow, depending on concentration.

3. **Mild Oxidizing Agents.** The agents KNO_2, H_2O_2, Fe^{3+}, $K_2Cr_2O_7$, and Cu^{2+} in dilute sulfuric acid solution oxidize I^- ion, but they do not readily react with bromide ion. With H_2O_2, Br^- interferes if its concentration is high; thiocyanates interfere with the Fe^{3+} reaction; with Cu^{2+} the reduction product is white, insoluble CuI:

$$2\,Cu^{2+} + 4\,I^- \longrightarrow I_2 + 2\,CuI(s)$$

4. **Strong Oxidizing Agents.** Cl_2, $KMnO_4$, and concentrated H_2SO_4 all liberate I_2 from iodide solutions, and they also oxidize bromides. If chlorine is used, an excess should be avoided, because the I_2 may be further oxidized to colorless IO_3^-:

$$Cl_2 + 2\,I^- \longrightarrow 2\,Cl^- + I_2$$
$$5\,Cl_2 + I_2 + 6\,H_2O \longrightarrow 10\,Cl^- + 2\,IO_3^- + 12\,H^+$$

THIOCYANATE, SCN⁻

Thiocyanate salts are all soluble in water, with the exception of those of the Group I cations and copper(I). The thiocyanates resemble the halides in many of their physical properties. Thiocyanates are subject to decomposition by both oxidizing and reducing agents.

CHARACTERISTIC PROPERTIES OF THE THIOCYANATE ION

1. **Silver Nitrate.** Mixing of solutions containing SCN^- and Ag^+ ions produces a white, curdy precipitate of AgSCN, which is soluble in ammonia but not in mineral acids. Boiling AgSCN with 1M NaCl converts the salt to AgCl and frees the SCN^- ion:

$$AgSCN(s) + Cl^- \rightleftharpoons AgCl(s) + SCN^-$$

2. **Ferric Nitrate.** A deep red complex ion, $FeSCN^{2+}$, forms on addition of Fe(III) solutions to thiocyanates under slightly acidic conditions. This is the classic test for the presence of thiocyanate ion in solution.

3. **Nitric Acid or Sulfuric Acid and Oxidizing Agents.** Warm 6M HNO_3, or 6M H_2SO_4 in the presence of oxidizing agents, including $K_2S_2O_8$, decomposes thiocyanates to various products, depending on which acid is used. These products typically do not form insoluble precipitates with $AgNO_3$ solutions.

4. **Zinc.** In dilute H_2SO_4 solution, zinc reduces thiocyanates to H_2S:

$$Zn(s) + SCN^- + 3\,H^+ \longrightarrow Zn^{2+} + H_2S(g) + HCN(g)$$

SULFIDE, S²⁻

As we have seen in studying the metallic cations, most of the transition metal sulfides are insoluble in water. Many of them, of course, are soluble in 6M HCl, and except for HgS, all are oxidized by hot 6M HNO_3.

Hydrogen sulfide in water solution behaves as a very weak acid ($K_1 = 1 \times 10^{-7}$, $K_2 = 1 \times 10^{-15}$). Because of extensive hydrolysis, sulfide salts are quite alkaline. One molar Na_2S is about 90 per cent hydrolyzed, but even solutions of NaHS are distinctly alkaline. The odor of H_2S is that of rotten eggs, and, once experienced, it is unforgettable; it is readily detected, even over very basic sulfide solutions. Sulfides are easily oxidized, with free sulfur or sulfate ion being the common product.

CHARACTERISTIC REACTIONS OF THE SULFIDE ION

1. **Silver Nitrate.** A very insoluble black precipitate of Ag_2S is obtained. The Ag_2S is insoluble in ammonia, but it is readily oxidized by hot 6M HNO_3.

2. **Hydrochloric Acid.** Many sulfides liberate H_2S when treated with 6M HCl; a typical reaction is:

$$MnS(s) + 2\,H^+ \longrightarrow Mn^{2+} + H_2S(g)$$

The gas can be detected by its odor or by its ability to blacken lead acetate paper. Sulfides that do not dissolve in HCl can be made to give off H_2S by adding metallic zinc to the HCl-sulfide mixture; the zinc reduces the metallic cation in the sulfide and thereby frees the S^{2-} ion for reaction with H^+. If zinc is used, sulfites and thiocyanates also react to give off H_2S. The presence of sulfides is usually established in one of the preliminary tests for the anions, and not as part of the chloride group.

3. **Oxidizing Agents.** Hot 6M HNO_3, or 6M HCl plus $KClO_3$, or aqua regia oxidizes sulfides, producing no H_2S. The product typically contains free sulfur, but some sulfate is always formed.

4. **Lead Nitrate.** This reagent is usually used on filter paper and reacts with sulfide ion in solution or with H_2S gas to produce shiny black or brown PbS. To prepare sulfide-sensitive paper, add 6M NaOH to a few milliliters of 0.1M $Pb(NO_3)_2$ until the precipitate dissolves. Moisten a piece of filter paper with the solution and use the paper to make the test.

PROCEDURE FOR ANALYSIS OF
ANIONS IN THE CHLORIDE GROUP

Because of interferences, one cannot in general use spot tests for identification of the anions in the Chloride Group, but must resort to a systematic procedure analogous to that used with the metallic cations. For this group you may assume, unless directed otherwise, that 5 ml of sample solution contains about 1 ml of 0.1M solutions of the sodium salts of one or more of the anions in the chloride group.

Step 1. To 3 ml of sample solution in a 30 ml beaker add 2 ml of water and 0.5 ml of 6M acetic acid. Boil the solution gently for about a minute. Then add about 0.1 g of solid potassium peroxydisulfate, $K_2S_2O_8$, and 0.5 ml of 6M acetic acid, and heat the liquid, just to the boiling point. Swirl the liquid occasionally. If the solution takes on a yellow color, which deepens with time, iodide is probably present. When the liquid just begins to boil, remove the beaker from the heat and put it on the bench top.

Step 2. *Confirmation of the presence of iodide.* Pour 1 ml of the solution in the beaker into a test tube, add 1 ml of CCl_4, and shake thoroughly to extract any iodine into the (lower) CCl_4 layer. If this layer is violet on settling, iodide is present in the sample.

Step 3. *Confirmation of the presence of thiocyanate.* To the (upper) water layer in the test tube from Step 2, add 1 or 2 drops of 0.1M $FeCl_3$; do not shake. A deep red coloration due to formation of $FeSCN^{2+}$ is proof of the presence of thiocyanate.

Step 4. Heat the rest of the solution from Step 1 to boiling, and boil it gently to drive out dissolved I_2. This takes time, and when the volume of liquid decreases by about one third, add 0.5 ml of 6M acetic acid, 0.1 g of $K_2S_2O_8$, and enough water to reestablish the original volume. Repeat this operation as necessary until the solution becomes essentially colorless.

Step 5. Remove the beaker from the heat, and add 0.5 ml of 6M H_2SO_4 and 0.1 g of $K_2S_2O_8$. Heat the liquid to boiling, but do not boil it. If bromide ion is present, and iodide is also, you will probably observe a very slight initial yellow coloration, due to residual iodide being oxidized, and, as the solution comes to the boil, liberation of Br_2, which gives the solution a yellow color. Remove the beaker from the heat.

Step 6. *Confirmation of the presence of bromide.* Pour 0.5 ml of the liquid in the beaker from Step 5 into a test tube, add 0.5 ml of CCl_4, and shake thoroughly. A yellow or orange color in the CCl_4 layer confirms the presence of bromide.

Step 7. *Confirmation of the presence of chloride.* Put the beaker with the solution from Step 5 back on the heater and gently boil the solution until the yellow color has been dispelled. Add a little $K_2S_2O_8$ to make sure that you have oxidized all the bromide ion, and, if necessary, boil again to drive out the liberated Br_2. To the clear liquid add 0.5 ml of 6M HNO_3. Mix the solution and add 1 or 2 drops of 0.1M $AgNO_3$. A white precipitate of AgCl confirms that chloride is in the sample.

Step 8. *Confirmation of the presence of sulfide.* If sulfide is present in the sample, it is detected in the preliminary test with acid. Sulfide combines with H^+ ion to form H_2S, with its impressive odor. Actually, if sulfide is present, you will probably be able to smell H_2S above the sample when you receive it, because hydrolysis of sulfides to H_2S is extensive even in moderately alkaline solutions.

To further confirm the presence of sulfide, pour the acidified solution from Preliminary Test 1 into a 30 ml beaker. Prepare a small strip of lead nitrate paper as directed under reactions of sulfide ion. Remove the excess liquid by touching the strip to a piece of dry filter paper, and then put the strip on the bottom of a small watch glass. Cover the beaker with the watch glass and heat it gently. If sulfide is present, the paper turns shiny black or brown as PbS is formed by reaction of Pb^{2+} with the H_2S.

COMMENTS OF THE PROCEDURE FOR
ANALYSIS OF THE CHLORIDE GROUP

Step 1. Boiling under the mildly acidic conditions removes most of the anions derived from volatile acids, and in particular drives out sulfides and sulfites, which are subject to oxidation in the later steps. If a precipitate forms on initial boiling, centrifuge it out before proceeding; the precipitate is probably sulfur. If iodide is present, it is nearly all oxidized by the time the solution of $K_2S_2O_8$ reaches the point of simmering. Do not boil the solution, because thiocyanate, which we are also trying to detect, is destroyed by any appreciable amount of boiling.

Step 2. The purple color of iodine in CCl_4 offers sensitive and conclusive evidence for the presence of iodides. No interferences are likely.

Step 3. Both thiocyanates and iodides give a reddish coloration with Fe^{3+}. The iodide is oxidized by Fe^{3+} to iodine, which probably gives most of the color; the SCN^- ion forms the very characteristic deep red $FeSCN^{2+}$ complex. To make the test for SCN^-, just let drops of the $FeCl_3$ solution fall into the water layer from Step 2; let the CCl_4 layer just stay there; it does not interfere. There is a negligible amount of iodide left at this point, and the appearance of a red color, which is usually quite deep, is a good test for thiocyanate.

Steps 4 and 5. The iodine is somewhat slow in boiling out, but it goes eventually. The advantage of using the peroxydisulfate is that it has no effect on bromides under the mildly acid conditions that prevail. On addition of the H_2SO_4, a small amount of residual I_2 is quickly oxidized; if bromides are there, they oxidize at boiling temperature, but not just in the warm solution. Since Br_2 is quite volatile, it is important that you do not boil the solution before making the test for bromide.

Step 6. If bromide is present, the color is yellow or orange. A violet color is just residual I_2. In general, there seems to be no difficulty in detecting bromides, but you have to remember that violet is not yellow or orange.

Step 7. If thiocyanates are present, they are oxidized by the boiling process, mainly in Step 4, and therefore they do not interfere here. The main possible difficulty is that the acidic solution is boiled down to too small a volume and HCl is driven out. This should not occur if you follow directions, but losing chloride by evolution of HCl is possible. If you are worried that the precipitate you obtain might be AgBr, centrifuge it out and try to dissolve it in 1 ml of 6M NH_3. Silver chloride dissolves readily; AgBr is essentially insoluble.

Step 8. Sulfides are really hard to miss, once you know what H_2S smells like, and what good chemistry student does not?

Outline of Procedure for Analysis of Anions in the Chloride Group

Ions possibly present:

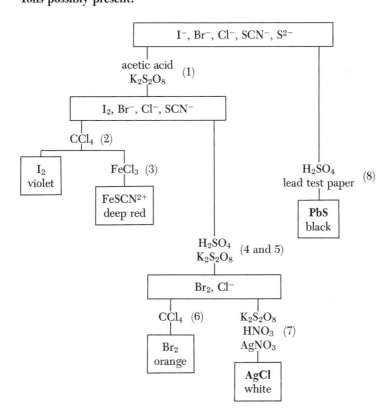

PROBLEMS

1. Write balanced net ionic equations to explain the following observations:
   a. Concentrated sulfuric acid reacts with a concentrated solution of NaCl to evolve a gas.
   b. A precipitate forms when solutions of $AgNO_3$ and NaBr are mixed; this precipitate dissolves in NaCN solution (two reactions).
   c. A precipitate forms when solutions of $AgNO_3$ and Na_2S are mixed.
   d. A piece of filter paper saturated with lead nitrate turns black when exposed to H_2S.

2. Complete and balance the following oxidation-reduction equations:
   a. $BiO_3^- + Cl^- + H^+ \longrightarrow Bi^{3+} + Cl_2(g) + H_2O$
   b. $PbBr_2(s) + NO_3^- + H^+ \longrightarrow Pb^{2+} + Br_2 + NO(g) + H_2O$
   c. $Br^- + SO_4^{2-} \longrightarrow SO_2(g)$ (acidic solution)
   d. $H_2O_2 + I^- \longrightarrow H_2O$ (acidic solution)
   e. $SO_4^{2-} + I^- \longrightarrow H_2S(g) + I_2$ (acidic solution)

3. Give the formula of a reagent that will oxidize:
   a. Br^- but not Cl^- c. S^{2-} but not Cl^-
   b. I^- but not Br^- d. SCN^- but not Cl^-

4. Describe a simple test to distinguish between the following:
   a. SCN^- and S^{2-} c. Cl^- and CO_3^{2-}
   b. Cl^- and Br^- d. Br^- and SO_4^{2-}

5. Explain why:
   a. In Step 1, the solution is boiled *before* adding $K_2S_2O_8$.
   b. In Step 1, the solution should not be boiled *after* adding $K_2S_2O_8$.
   c. The solution is boiled in Step 4 but not in Step 5.

6. Describe exactly what would happen in each step of the chloride group analysis with unknowns containing the following:
   a. Cl^-, I^-, and SCN^- b. Br^-, I^-, and S^{2-}

7. Devise a simple scheme of analysis, including no extra steps, for an unknown containing only the following anions:
   a. Cl^-, SCN^-, and S^{2-} b. SO_4^{2-}, SO_3^{2-}, and SCN^-

8. An unknown belonging to the Chloride Group is analyzed with the following results:

   The unknown, when heated with $K_2S_2O_8$ in acetic acid solution, undergoes no color change. However, when H_2SO_4 is added and the solution is reheated, a yellow color appears. Upon prolonged boiling the yellow color disappears; addition of $AgNO_3$ to this solution gives a precipitate.

   Which ions are present? Which are absent? Which are questionable?

9. An unknown may contain no anions except CO_3^{2-}, SO_4^{2-}, Cl^-, and I^-. Addition of dilute H_2SO_4 to a sample of the unknown causes evolution of a gas; when $BaCl_2$ is added to the acidic solution, a white precipitate forms. When $K_2S_2O_8$ is added to another portion of the acidic solution, a yellow color appears

on heating. This solution is boiled and treated with $AgNO_3$ to give a white precipitate.

Which ions are present? Which are absent? Which are questionable?

10. Devise a system of analysis for a solid that may contain $PbCrO_4$, Na_2SO_3, $NaCl$, and AgI.

101. Write balanced net ionic equations for the reactions that occur when:
   a. Silver chloride is exposed to a solution of sodium sulfide.
   b. Solutions of $AgNO_3$ and $NaSCN$ are mixed.
   c. Ferrous sulfide is treated with concentrated HCl.

102. Write balanced net ionic equations to represent the oxidation-reduction reactions that occur when:
   a. Solid silver bromide is treated with zinc and acid.
   b. Concentrated nitric acid is added to a solution of sodium iodide.
   c. Manganese dioxide reacts with concentrated HBr.
   d. Potassium peroxydisulfate reacts with I^- ions.
   e. Zinc is added to an acidified solution of NaSCN.

103. Which of the ions in the chloride group are readily oxidized by the following:
   a. Concentrated H_2SO_4?   c. $K_2S_2O_8$ in H_2SO_4?
   b. Concentrated HNO_3?   d. $K_2S_2O_8$ in acetic acid?

104. Describe a simple test for distinguishing between the following:
   a. I^- and Br^- c. CrO_4^{2-} and I^-
   b. Cl^- and S^{2-} d. SO_3^{2-} and SCN^-

105. Explain why:
   a. Acetic acid rather than sulfuric acid is used in Step 1.
   b. A red color may be obtained in Step 3 in the absence of SCN^-.
   c. The test for Cl^- (Step 7) cannot be carried out on the original solution.

106. Describe exactly what would happen in each step of the chloride group analysis with unknowns containing the following:
   a. Cl^-, Br^-, and I^- b. I^-, SCN^-, and S^{2-}

107. Devise a simple scheme of analysis, including no extra steps, for an unknown containing no anions other than the following:
   a. Cl^-, Br^-, and SCN^- b. CO_3^{2-}, CrO_4^{2-}, and I^-

108. An unknown in the chloride group is analyzed with the following results.

Addition of H_2SO_4 to the solution causes evolution of a gas with a foul odor. A little $K_2S_2O_8$ is then added and the solution is heated to give a yellow color. Upon prolonged heating, the solution becomes colorless and fails to give a precipitate with $AgNO_3$.

Which ions are present? Which are absent? Which are questionable?

109. An unknown may contain no anions other than CrO_4^{2-}, SO_3^{2-}, Br^-, SCN^-, and S^{2-}. A sample of the unknown is heated with HCl; no gas is evolved. Addition of $AgNO_3$ to a portion of the acidic solution produces a precipitate; addition

of $BaCl_2$ to another portion of the solution fails to produce a precipitate. When a sample of the unknown is treated with $FeCl_3$, a deep red color appears.

Which ions are present? Which are absent? Which are questionable?

110. Devise a system of analysis for a solid that may contain CaC_2O_4, $AgBr$, $KSCN$, and CuS.

LABORATORY ASSIGNMENTS

1. Make up a known sample containing the anions in the Chloride Group. Go through the standard procedure for analysis with that sample, noting your observations and comparing them with those predicted in the scheme. Obtain a Chloride Group unknown from your instructor and analyze it by the procedure. Indicate your observations and conclusions on a flow chart.

2. Given a sample known to contain only two anions, selected from the Chloride Group, develop a scheme to determine which anions are present. Draw a flow chart summarizing your approach, noting all reagents to be used and the species expected at each step, depending on which anions are present. When you are sure your procedure will work, use it to analyze an unknown your instructor will furnish you. Record your observations and conclusions on the flow chart and submit it for evaluation. Your instructor may then issue you another unknown for analysis.

3. You will be given a sample that contains two anions, one from the Sulfate Group and one from the Chloride Group. Use the preliminary tests, the standard procedures for the group, or your own scheme to determine which anions are present. Since only one anion from the Chloride Group can be present, it is possible to set up a very simple procedure for analysis for that anion. Report your observations and conclusions to your instructor, who may issue you another sample for analysis.

4. You will be given a sample that may contain as many as four anions, chosen from one of the following sets. When you are assigned a set of anions, develop a procedure by which you can analyze a sample for those anions only. Draw a flow chart for your scheme, indicating all reagents to be used and species that may be formed in each step. Test your method with a known sample, and when you are sure it works, analyze an unknown sample that your instructor will furnish. Report your observations and conclusions on the flow chart. Your set of anions will be one of the following:

    a. Cl^-, SCN^-, CO_3^{2-}, and SO_3^{2-} e. I^-, SCN^-, $C_2O_4^{2-}$, and SO_3^{2-}
    b. Br^-, S^{2-}, PO_4^{3-}, and $C_2O_4^{2-}$ f. Br^-, SCN^-, SO_3^{2-}, and PO_4^{3-}
    c. Br^-, I^-, SO_4^{2-}, and SO_3^{2-} g. Cl^-, Br^-, $C_2O_4^{2-}$, and CO_3^{2-}
    d. Cl^-, I^-, PO_4^{3-}, and SO_4^{2-} h. S^{2-}, SCN^-, SO_4^{2-}, and CO_3^{2-}

SECTION 4

THE PROPERTIES OF THE ANIONS IN THE NITRATE GROUP—NO_3^-, NO_2^-, $C_2H_3O_2^-$, ClO_3^-

NITRATE, NO_3^-

Since essentially all nitrate salts are soluble in water, they are very commonly used to furnish required cations in laboratory stock solutions. In some cases, essentially only the nitrate salt of a particular cation is commercially available.

Nitric acid, HNO_3, from which nitrate salts can be derived, can be prepared in anhydrous form, and it is a liquid boiling at about $86°$ C. The commercial solution is about 15M, and is a constant boiling mixture ($120°$ C), which means that the boiling liquid and its vapor have the same composition. This makes HNO_3 a reasonably volatile acid, similar to HCl, but unlike H_2SO_4 and H_3PO_4, which are not readily boiled out of solution. Hot 6M HNO_3 is a good oxidizing agent as well as a strong acid. Consequently it is a good solvent for many metals and nearly all the sulfides, as well as hydroxides and the salts of weak acids.

In general, nitrates are not as thermally stable as the halides or the sulfates. Most of the heavy metal nitrates decompose to the oxides when strongly ignited:

$$2\,Cu(NO_3)_2(s) \longrightarrow 2\,CuO(s) + 4\,NO_2(g) + O_2(g)$$

CHARACTERISTIC REACTIONS OF THE NITRATE ION

1. **Manganous Chloride.** Nitrate ion, and all oxidizing anions, react with concentrated Mn(II) solutions in strong HCl to form a dark brown or black complex of Mn(IV), presumably $MnCl_6^{2-}$:

$$3\,MnCl_4^{2-} + 2\,NO_3^- + 6\,Cl^- + 8\,H^+ \longrightarrow$$
$$3\,MnCl_6^{2-} + 2\,NO(g) + 4\,H_2O$$

2. **Potassium Iodide.** In cold dilute acid solution, nitrates do not react with iodides. Upon heating, I_2 is slowly formed and the solution turns yellow:

$$6\,I^- + 2\,NO_3^- + 8\,H^+ \longrightarrow 3\,I_2 + 2\,NO + 4\,H_2O$$

3. **Sulfuric Acid.** Hot 18M H_2SO_4 causes evolution of brown NO_2 gas from solid nitrates. In most cases, 6M H_2SO_4 is without effect on nitrate solutions. Clean copper turnings, added to a nitrate solution made strongly acidic with H_2SO_4, dissolve with evolution of brown NO_2 gas; nitrites exhibit the same reaction with copper.

4. **Ferrous Sulfate.** If a concentrated solution of $FeSO_4$ is brought into contact with a nitrate solution made highly acidic with H_2SO_4, the nitrate ion is reduced to NO and combines with excess Fe(II) to form an unstable brown complex ion, $Fe(NO)^{2+}$:

$$3\,Fe^{2+} + NO_3^- + 4\,H^+ \longrightarrow 3\,Fe^{3+} + NO + 2\,H_2O$$
$$Fe^{2+} + NO \rightleftharpoons Fe(NO)^{2+}$$

Unfortunately, nitrites, chromates, iodides, and bromides may interfere with this test.

5. **Aluminum or Zinc.** Finely divided aluminum or zinc particles in NaOH solution reduce nitrates to NH_3, which can then be detected by odor or by litmus paper. Nitrites and thiocyanates, and, of course, ammonium salts, interfere and must be removed before the test is carried out.

NITRITE, NO_2^-

Nitrite salts are typically soluble in water, the exception being $AgNO_2$, which is about 0.02M in its saturated solution at 25° C. The alkali nitrites, the usual source of the NO_2^- ion in the laboratory, are made by strongly heating the corresponding nitrates. If NO_2 is dissolved in water, a mixture of nitrous acid, HNO_2, and nitric acid is obtained:

$$2\,NO_2 + H_2O \rightleftharpoons HNO_2 + H^+ + NO_3^-$$

Nitrous acid is weak ($K = 4.5 \times 10^{-4}$) and decomposes on being heated; its solution is pale blue.

Nitrites can serve as both reducing and oxidizing agents, depending on conditions. The NO_2^- ion is also a good complexing agent. Because solutions of nitrites slowly decompose on standing, they should be freshly prepared.

CHARACTERISTIC REACTIONS OF THE NITRITE ION

1. **Potassium Iodide.** Nitrites oxidize iodides to iodine in cold, dilute sulfuric or acetic acid solution. Nitrates do not interfere.

$$2\,NO_2^- + 2\,I^- + 4\,H^+ \longrightarrow I_2 + 2\,NO + 2\,H_2O$$

2. **Ferrous Sulfate.** Nitrite ion in dilute acid solution oxidizes Fe^{2+} to Fe^{3+}. Excess Fe^{2+} reacts with the NO produced, forming a brown solution of $Fe(NO)^{2+}$. Nitrates participate in this reaction only in concentrated H_2SO_4.

3. **Sulfuric Acid.** Six molar H_2SO_4 decomposes nitrites with the evolution of brown NO_2 gas. The reaction with 18M H_2SO_4 is very rapid:

$$NO_2^- + H^+ \rightleftharpoons HNO_2$$
$$3\,HNO_2 \longrightarrow H^+ + NO_3^- + 2\,NO(g) + H_2O$$
$$2\,NO(g) + O_2(g) \longrightarrow 2\,NO_2(g)$$

4. **Urea.** In HCl solution, nitrites oxidize urea to CO_2 and are themselves reduced to nitrogen gas; nitrates do not interfere:

$$2\,NO_2^- + CO(NH_2)_2 + 2\,H^+ \longrightarrow 2\,N_2(g) + CO_2(g) + 3\,H_2O$$

5. **Ammonium Chloride.** Nitrites decompose when boiled in solution with ammonium salts, and thereby can be removed completely from a system:

$$NO_2^- + NH_4^+ \longrightarrow N_2(g) + 2\,H_2O$$

6. **Potassium Permanganate.** Acidic nitrite solutions decolorize this reagent, as do other reducing anions:

$$2\,MnO_4^- + 5\,NO_2^- + 6\,H^+ \longrightarrow 2\,Mn^{2+} + 5\,NO_3^- + 3\,H_2O$$

CHLORATE, ClO_3^-

Chlorate salts are generally soluble in water, but being strongly oxidizing, they are not usually used as sources of metallic cations. The most common chlorate salt is $KClO_3$, which is prepared by bubbling chlorine gas into warm, concentrated KOH solution:

$$3\,Cl_2 + 6\,OH^- \rightleftharpoons ClO_3^- + 5\,Cl^- + 3\,H_2O$$

Chloric acid, $HClO_3$, from which chlorates can be considered to be derived, cannot be isolated in pure form, because it decomposes spontaneously to ClO_2 and ClO_4^-. One must never add 18M H_2SO_4 to a solid chlorate, because an explosion results; this means that you must always test to see that a solid unknown does not contain chlorate before performing any test with concentrated sulfuric acid. It is important to make this test, because one of the things you do not need on a pleasant afternoon in the laboratory is to be sprayed with the mixture of 18M H_2SO_4 and ClO_2 that is ejected by the explosion just mentioned.

CHARACTERISTIC REACTIONS OF THE CHLORATE ION

1. **Potassium Iodide.** Chlorates in warm acid oxidize iodide ion to form a yellow or orange solution of I_2. The reaction is very slow at room temperature:

$$ClO_3^- + 6\,I^- + 6\,H^+ \longrightarrow Cl^- + 3\,I_2 + 3\,H_2O$$

2. **Sodium Nitrite.** This reagent reduces chlorates to chlorides, which may then be tested with $AgNO_3$. Acidic H_2S, or metallic iron, zinc, or aluminum, also serves for the reduction:

$$3\,NO_2^- + ClO_3^- \longrightarrow 3\,NO_3^- + Cl^-$$

3. **Hydrochloric Acid.** Hot 6M HCl is oxidized by chlorates, producing a yellow solution:

$$2\,ClO_3^- + 2\,Cl^- + 4\,H^+ \longrightarrow 2\,ClO_2(g) + Cl_2(g) + 2\,H_2O$$

The gases evolved are strong oxidants and easily bleach moist litmus paper.

4. **Soluble Sulfites.** A slightly acidic solution of Na_2SO_3 reduces chlorates to chlorides:

$$3\,(H_2SO_3) + ClO_3^- \longrightarrow 3\,HSO_4^- + Cl^- + 3\,H^+$$

ACETATE, $C_2H_3O_2^-$

Acetates are characteristically soluble in water and are certainly the most common of the salts of organic acids.

Acetic acid, $HC_2H_3O_2$, is a weak acid, which can lose one H^+ ion:

$$HC_2H_3O_2 \rightleftharpoons H^+ + C_2H_3O_2^- \qquad K = 1.8 \times 10^{-5}$$

The pure acid is available commercially as glacial acetic acid and is a liquid that boils at 118° C and freezes at 16.6° C. It has the characteristic odor of vinegar.

Acetates in solution resist oxidation, but if the solid salts are heated strongly, they decompose, leaving the carbonate or the oxide of the metal as a residue. There is often some charring during the ignition.

CHARACTERISTIC REACTIONS OF THE ACETATE ION

1. **Sulfuric Acid.** If an acetate is heated with sulfuric acid, acetic acid is volatilized and can be detected by its vinegar odor:

$$C_2H_3O_2^- + H^+ \rightleftharpoons HC_2H_3O_2$$

2. **Heat.** If a solid acetate is heated, it may darken or char and give off vapors with characteristic odors.

PROCEDURE FOR ANALYSIS OF ANIONS IN THE NITRATE GROUP

Unless told otherwise, you may assume that 5 ml of sample contains 1 ml of a 0.1M solution of the sodium salt of one or more of the anions in the nitrate group.

Step 1. *Test for the presence of nitrite.* Nitrite ion reacts with other easily oxidized or reduced anions, and therefore it tends to be present only in systems in which those ions are absent. To 1 ml of sample solution add 0.5 ml of 6M H_2SO_4. In a separate test tube, dissolve 0.1 g of $FeSO_4 \cdot$ 7 H_2O in 1 ml of water. Mix the two solutions. If nitrite ion is present, you will obtain a dark greenish brown solution of $Fe(NO)^{2+}$.

Step 2. *Test for the presence of chlorate.* To 1 ml of sample solution in a test tube add 2 ml of 0.1M $AgNO_3$. Mix well and put the test tube in the boiling water bath for a minute or two. Centrifuge and decant the liquid into a test tube; any precipitate may be discarded. Add 1 or 2 drops of 0.1M $AgNO_3$ to the liquid to make sure Ag^+ ion is present in excess. If it is not, add another ml of $AgNO_3$ and centrifuge out the solid that precipitates. To the clear decanted liquid add 0.5 ml of 6M HNO_3 and mix well. Then add 1 ml of 0.1M $NaNO_2$. If chlorate is present, it is reduced by the nitrite ion and a white precipitate of AgCl forms gradually.

Step 3. *Test for the presence of nitrate.* To 1 ml of sample solution *carefully* add 2 ml of 18M H_2SO_4; because this is concentrated acid, add the acid slowly, mixing all the while. Cool the hot test tube under the water tap. In another test tube dissolve about 0.1 g of $FeSO_4 \cdot 7$ H_2O in 1 ml of water. Holding the tube containing the H_2SO_4 at an angle of about 45 degrees, let 5 drops of the $FeSO_4$ solution from a medicine dropper run down the side of the tube and form a layer over the acid. Let the test tube stand for a few minutes, and look for a brown ring at the junction of the two layers. In the absence of interfering anions, this ring is proof of the presence of nitrate ion. Nitrites, chromates, iodides, and bromides may interfere with this test and should be removed before it is performed.

Step 4. *Test for the presence of acetate.* Acetate ion is detected in Preliminary Test 1 for volatile acids. If acetate is present, acetic acid forms on addition of H_2SO_4, and the solution smells like vinegar. There is really no better test for acetate than this, because acetate forms no common insoluble salts and is not subject to easy oxidation or reduction. In the absence of other anions that form volatile acids, the test for acetate is adequate. In the presence of sulfides or sulfites, it is not easy to prove that acetate is also there.

COMMENTS ON PROCEDURE FOR ANALYSIS
OF THE NITRATE GROUP

Step 1. Nitrites do not coexist well with ions that are subject to either oxidation or reduction; if nitrite is present, many other anions are not, and conversely. The anions that react with nitrite include CrO_4^{2-}, ClO_3^-, I^-, Br^-, SO_3^{2-}, S^{2-}, and SCN^-. In this test, some of these ions produce colored solutions (chromates turn blue-green, thiocyanates turn orange), but the $Fe(NO)^{2+}$ complex is dark and quite distinctive, and therefore there should be no real problem with interferences. If you are in doubt, try the test with a solution of 0.1M $NaNO_2$.

Step 2. Here it is essential that Ag^+ ion be present in excess, so that any Cl^- ion produced by chlorate reduction precipitates. Silver nitrite is not very soluble, but it does not tend to precipitate under the conditions prevailing.

Step 3. The problem with any test for nitrate is the likely interferences. Both bromides and iodides are oxidized by strong H_2SO_4, giving dark colored solutions. Nitrites produce the same ring as nitrates, only more so. Chromates are reduced by Fe^{2+}, producing a nice green ring, possibly superposed on the brown one. One advantage is that the anions that interfere do not in general coexist, so that you can have iodide and/or bromide, or nitrite, or chromate, as an interfering anion.

Iodide and bromide can be easily removed by adding an excess of a solution of Ag_2SO_4 in 1M H_2SO_4, centrifuging out the solid, and proceeding to test the liquid as directed. Chromate is precipitated readily by excess 0.1M $Pb(C_2H_3O_2)_2$ and can then be centrifuged out. Nitrite decomposes, producing N_2 gas as the only product, if one adds 1 ml of 2M NH_4Cl to the solution and boils it gently for a minute.

Having removed the interfering anions, you then can get a meaningful result in the nitrate test. The test is a good one and is quite sensitive. It is best viewed against white paper.

PROBLEMS

1. Write balanced net ionic equations for the oxidation of I^- in acidic solution by the following:
   a. NO_3^- b. NO_2^- (assume reduction to NO)   c. ClO_3^- (assume reduction to Cl^-)

2. List as many anions as you can think of that could not be present, in acidic solution, with the following:
   a. NO_2^- b. ClO_3^- c. CO_3^{2-}

3. How would you distinguish between the following:
   a. $C_2H_3O_2^-$ and NO_2^-?   c. ClO_3^- and NO_3^-?
   b. $C_2H_3O_2^-$ and NO_3^-?   d. ClO_3^- and CrO_4^{2-}?

4. How would you accomplish the following conversions?
   a. $C_2H_3O_3^- \longrightarrow HC_2H_3O_2$ d. $NO_3^- \longrightarrow NO_2(g)$
   b. $NO_2^- \longrightarrow NO_3^-$ e. $MnO_4^- \longrightarrow Mn^{2+}$
   c. $ClO_3^- \longrightarrow Cl^-$ f. $Br^- \longrightarrow Br_2$

5. How would you remove the following ions from water solution?
   a. ClO_3^- b. Cl^- c. SO_3^{2-} d. S^{2-}

6. What, if anything, happens to the NO_2^- ion in each step of the following tests?
   a. The preliminary tests   c. The Cl^- group tests
   b. The SO_4^{2-} group tests   d. The NO_3^- group tests

7. An unknown that may contain no anions other than CrO_4^{2-}, SO_3^{2-}, I^-, NO_3^-, and NO_2^- is analyzed with the following results.
   Addition of dilute H_2SO_4 followed by $FeSO_4$ gives a dark greenish brown solution. With concentrated H_2SO_4 and $FeSO_4$, a very faint brown ring appears. Which ions are present? Which are absent? Which are questionable?

8. Devise a system of analysis for an unknown that may contain no anions other than SO_3^{2-}, SO_4^{2-}, I^-, Cl^-, and NO_3^-.

101. Write balanced net ionic equations for the following reactions:
   a. The reduction of NO_2^- to $NO(g)$ by Fe^{2+} (acidic solution).
   b. The reduction of NO_2^- to N_2 by NH_4^+.
   c. The oxidation of NO_2^- to NO_3^- by MnO_4^- (acidic solution).
   d. The oxidation of NO_2^- to NO_3^- by $Cr_2O_7^{2-}$ (acidic solution).

102. Which of the following pairs of ions would you *not* expect to find together in acidic solution?
   a. $C_2H_3O_2^-$ and ClO_3^- d. ClO_3^- and I^-
   b. NO_2^- and ClO_3^- e. ClO_3^- and SO_3^{2-}
   c. NO_2^- and NO_3^-

103. How would you distinguish between the following:
   a. ClO_3^- and Cl^-?   c. NO_3^- and NO_2^-?
   b. $C_2H_3O_2^-$ and CO_3^{2-}?   d. NO_3^- and SO_4^{2-}?

104. How would you accomplish the following conversions?
    a. $Cl^- \longrightarrow Cl_2(g)$ d. $Fe^{2+} \longrightarrow Fe^{3+}$
    b. $HC_2H_3O_2 \longrightarrow C_2H_3O_2^-$ e. $AgClO_3(s) \longrightarrow AgCl(s)$
    c. $NO_2^- \longrightarrow N_2(g)$

105. How would you remove the following ions from water solution?
    a. I^- b. NO_2^- c. Br^- d. CO_3^{2-}

106. What, precisely, happens to the ClO_3^- ion in each step of the following tests:
    a. The preliminary tests?   c. The Cl^- group tests?
    b. The SO_4^{2-} group tests?   d. The NO_3^- group tests?

107. An unknown that may contain no anions other than CO_3^{2-}, Cl^-, NO_3^-, $C_2H_3O_2^-$, and ClO_3^- is analyzed with the following results.
    When dilute H_2SO_4 is added to a sample of the unknown, it does not evolve a gas but does produce a sharp odor. When $AgNO_3$ is added to another sample of the unknown that has been acidified with HNO_3, it fails to produce a precipitate until $NaNO_2$ is added, at which point a white precipitate forms.
    Which ions are present? Which are absent? Which are questionable?

108. Devise a scheme of analysis for an unknown that may contain no anions other than CrO_4^{2-}, SCN^-, S^{2-}, ClO_3^-, and $C_2H_3O_2^-$.

LABORATORY ASSIGNMENTS

1. Prepare a known sample containing chlorate, nitrate, and acetate ions at about 0.1M concentrations. Test the sample for each of these anions, noting your observations and comparing them with those described in the procedure. Test a separate sample containing 1 ml of 0.1M KNO_2 for the presence of nitrite ion. Obtain an unknown from your instructor and analyze it for the possible presence of NO_2^-, ClO_3^-, NO_3^-, and $C_2H_3O_2^-$. When you believe you know the composition of the unknown, check your results by making a known of the same composition and testing it to see that it behaves as expected.

2. You will be given a sample that may contain anions included in one of the following sets. When you have been assigned a set, develop a procedure for analysis of a solution in which those anions may be present. Draw a flow chart for your procedure, and check to see that your system works by applying it to a known sample. When you are sure that you have a viable procedure, obtain an unknown from your instructor and analyze it for the presence of the anions in your set. The sets to be considered include the following:

         a. NO_3^-, ClO_3^-, Cl^-, SO_4^{2-}, and I^-
         b. ClO_3^-, $C_2H_3O_2^-$, SCN^-, CO_3^{2-}, and PO_4^{3-}
         c. NO_3^-, NO_2^-, SO_4^{2-}, I^-, and Cl^-
         d. ClO_3^-, CrO_4^{2-}, $C_2O_4^{2-}$, CO_3^{2-}, and $C_2H_3O_2^-$
         e. NO_3^-, $C_2H_3O_2^-$, SCN^-, SO_3^{2-}, and SO_4^{2-}
         f. NO_2^-, Cl^-, Br^-, PO_4^{3-}, and $C_2O_4^{2-}$
         g. NO_3^-, Br^-, SCN^-, SO_4^{2-}, and CO_3^{2-}
         h. $C_2H_3O_2^-$, I^-, CrO_4^{2-}, $C_2O_4^{2-}$, and PO_4^{3-}

3. You will be given a sample containing two anions chosen from any of the groups. Before coming to laboratory, outline how you will proceed to determine which anions are present. Apply your procedure to the unknown furnished to you. When you feel that you know the composition of the unknown, make up a sample containing the two anions you think are present and test that sample to make sure that it behaves in the same manner as your unknown. Submit your observations and conclusions to your instructor for evaluation. You may then be issued another unknown containing two anions for analysis.

CHAPTER 5

PREPARATION OF SOLID SAMPLES FOR ANALYSIS

So far in our work we have dealt with solutions of the ions under consideration, assuming in general that the cations were in the presence of nitrate or chloride ions and that the anions were obtained from solutions of their sodium or potassium salts. It is by no means necessary to limit qualitative analyses to such systems, and in particular it is important to consider how one would proceed to identify the cations and anions present in solid substances. Such solids might be pure metals, alloys, pure salts or oxides, or mixtures of these substances. The general problem in working with solids is getting them into solution, so that the procedures we have been using for analysis of cations and anions can be applied.

In many cases the preparation of solid samples for analysis is far from trivial. Minerals may have been solidified at high temperatures eons ago and may be very reluctant to dissolve even in concentrated reagents. In addition, minerals may be complex silicates, with properties that require special attention beyond the level of this course. Of necessity we restrict our attention to alloys composed of the metals whose cations we have studied, or salts of those cations and the anions we have investigated. Even these limitations will give us plenty to do, but it must be admitted that we are avoiding some of the more difficult possibilities. Since alloys are somewhat easier to handle than are the salts, we shall deal with them first.

TREATMENT OF ALLOYS

When a molten metallic solution is crystallized, the solid obtained may be a solid solution, or a mixture of small crystals of pure metals, or it may consist of one or more intermetallic compounds of fixed composition; in any case, the metallic solid is called an alloy.

On dissolving an alloy one obtains a solution of the cations of the metallic elements in the alloy. Analysis of the alloy then consists of determining the cations in its solution.

Dissolving a metal in aqueous solvents always involves its oxidation. With active metals, such as zinc and aluminum, solution may occur readily in

6M HCl, with H^+ ion being reduced to H_2 gas. Less electropositive metals, such as copper or mercury, dissolve only in an oxidizing acid, with 6M HNO_3 being quite effective in most cases. Metals verging on being nonmetallic, such as antimony, exist in solution mainly as complex anions, and these may dissolve only in a mixture of concentrated hydrochloric and nitric acids, the reagent we call aqua regia.

Although you might guess that aqua regia would be the reagent of choice, because it should dissolve any metal, this does not necessarily follow, since some of the rather active metals, such as aluminum and chromium, tend to become passive in concentrated HNO_3. Our procedure for dissolving alloys takes into account these factors, and is designed for use of the least concentrated reagent possible to accomplish the solution. We first try 6M HCl as a solvent. If that fails, we proceed with 12M HCl, then a mixture of 12M HCl and 6M HNO_3, and finally, if necessary, aqua regia, a mixture of 12M HCl and 17M HNO_3. You can be sure that this approach allows you to dissolve any alloy you may encounter, and also gives you information about the likely composition of the alloy.

PROCEDURE FOR DISSOLVING
AN ALLOY

Unless directed otherwise, you should use about 0.25 g of a finely divided sample of alloy. If the metal is in large chunks, you should file it, saw it, or otherwise reduce the size of the particles if possible, because solution occurs much more rapidly with powdered or granular metals.

Step 1. Put the sample of alloy in a 30 ml beaker and add 5 ml of 6M HCl. If a reaction begins immediately, let it proceed until it is complete, adding 6M HCl if it appears to slow down. If nothing occurs initially, heat the acid gently to boiling and let it simmer until the volume becomes about 2 ml. If the metal seems to respond to hot 6M HCl, add 3 ml more and continue simmering until solution is complete or no further reaction occurs. If hot 6M HCl is without effect, or does not completely dissolve the alloy, repeat the simmering with 12M HCl, moving your operation to the hood. If a still stronger reagent is needed, add 2 ml of 12M HCl and 2 ml of 6M HNO_3 to the 2 ml of remaining acid, and gently simmer that solution to a volume of 1 ml. As a last resort, try using 2 ml of 12M HCl and 2 ml of 17M HNO_3. You may stop any of these approaches as soon as all the alloy is dissolved.

When solution appears to be essentially complete, boil the liquid down to a volume of 1 ml, and then add 5 ml of water, mixing it well into the solution to dissolve any crystals of salt on the walls of the beaker. A white precipitate may appear on addition of water; it may contain $AgCl$, $PbCl_2$, $BiOCl$, and $SbOCl$. Stir the solution for a minute and then transfer it to a test tube. Centrifuge out any solid precipitate and decant the liquid into a small flask. The liquid may contain cations from Groups II to IV and is an acidic ($[H^+] \cong 1M$) solution of the nitrate and chloride salts of those cations; their concentrations should be about 0.1M.

You may proceed to analyze that solution for cations by the Procedures for Analysis of Groups II to IV. If you wish, you may try some preliminary tests, involving treatment of 0.5 ml portions of the solution with 6M NaOH, 6M NH_3, or 6M Na_2CO_3 to try to establish the presence of cations that form complex ions with NH_3 or OH^-. Such preliminary tests may be very helpful in furnishing clues as to which cations are likely to be present and may simplify your use of the Procedures for Analysis considerably. Additional information may be furnished if the solution has a color characteristic of that of a given cation (Table 5-1).

Step 2. Any precipitate remaining after the treatment with water should be washed twice with 3 ml of water; centrifuge out the solid each time and discard the wash.

Check for the presence of silver and lead in the precipitate by carrying out Steps 2 to 5 in the Procedure for Analysis of Group I Cations. You will

TABLE 5-1.   **Colors of Ions in Aqueous Solution**

Ion	Color	Ion	Color
Cu^{2+}	blue	Fe^{2+}	pale green
Ni^{2+}	green	Fe^{3+}	orange to yellow
Co^{2+}	pink	Cr^{3+}	bluish gray
Mn^{2+}	pale pink	CrO_4^{2-}	yellow
		$Cr_2O_7^{2-}$	orange

The colors of some of the cations may change considerably in the presence of complexing ligands.

not obtain a test for mercury(I), because any mercury present in the alloy would have been oxidized to Hg(II) and would appear in the Group II analysis. Antimony and bismuth, if present, do not interfere.

Step 3. If a precipitate remains after Step 4 in the Group I Procedure, it may contain the oxychlorides of bismuth or antimony. Wash the precipitate with 3 ml of water, centrifuge, and discard the wash. Add 1 ml of 6M NaOH to the precipitate, and stir well to dissolve any Sb(III) present. Centrifuge out any remaining solid, and decant the solution into a test tube. The solid may be $Bi(OH)_3$, which may be confirmed by carrying out the last part of Step 7 in the Group II Procedure. The liquid may contain $Sb(OH)_4^-$. The presence of antimony can be confirmed by acidifying the solution with 6M HCl, bringing the pH to 0.5, and continuing them with Step 18 in the Group II Procedure.

COMMENTS ON PROCEDURE FOR DISSOLVING AN ALLOY

Step 1. The response of alloys to the acid solvents varies tremendously with the nature of the alloy. Those that contain aluminum, zinc, or magnesium as their main components react very readily with 6M HCl, leaving only small amounts of or no residue. The ferrous alloys, containing perhaps chromium, manganese, cobalt, or nickel, in addition to iron, probably require at least 6M HNO_3 for complete reaction. Brasses and bronzes may be attacked by 12M HCl but also require 6M HNO_3 to dissolve all the alloy. Antimony, bismuth, and tin dissolve best in aqua regia. Depending on the degree of reaction with the various acid mixtures, you can often draw some very helpful conclusions about the general nature of the alloy.

In the concentrated acid mixtures, even Ag^+ and Pb^{2+} from chloro complex ions (e.g., $AgCl_2^-$, $PbCl_4^{2-}$) and are soluble. Dilution of the solution precipitates the chlorides, AgCl and $PbCl_2$, and may also cause some BiOCl and SbOCl to precipitate. Tests for Bi^{3+} and Sb^{3+} may also be obtained during analysis for Group II cations, since only partial precipitation of those cations tends to occur in this step.

One difficulty in analysis of alloys is that they are typically not mixtures of approximately equal amounts of the component metals, but rather they contain large percentages of one or two metals and small or very small amounts of other metals. The methods we have used can in general detect about 1 mg of metallic cation in 1 ml of solution; an alloy sample weighing 0.25 g should allow you to detect satisfactorily most metals if they are present at a concentration of 1 per cent or greater by weight. Your problem may be that, given a very convincing test for one cation, you may be unwilling to accept a relatively weak test for another. Alloys, however, frequently afford just that sort of situation. If you do obtain indication of a small amount of a given cation, you might profitably set up a procedure specifically designed to determine whether that cation is really present.

Step 2.   Silver and lead can be detected very well in the presence of BiOCl and SbOCl, neither of which is very soluble in hot water or 6M NH_3.

Step 3.   In many ways, analysis for bismuth and antimony is made more easily by this method than by the standard Group II procedure. The cation Sb(III) does not interfere with the reduction of Bi(III) by Sn(II) in the alkaline solution. Since bismuth is not amphoteric, the separation of Sb(III) as the hydroxo complex is a natural method to use.

TREATMENT OF NONMETALLIC SOLIDS
AND GENERAL MIXTURES

When a solid sample may contain salts or oxides, plus possibly alloys, the procedure of analysis must be modified from that used for metallic samples. The general solid may contain anions, which must be identified, and may also contain substances that do not dissolve in any of the common reagents.

With such samples our approach is to make preliminary solubility tests with various reagents, since these may tell us a great deal about the solid, at least in some cases. Using this preliminary information, we will dissolve as much of the solid as possible to form a solution that we can use for analysis of the cations it contains. Our reagents are less concentrated than are those used with some of the alloys, and in some cases we may obtain insoluble residues, which will be handled separately. The next step is to carry out preliminary tests for the determination of the anions present in the solid, which helps us to decide which specific anion tests to carry out.

PROCEDURE FOR ANALYSIS OF
SOLIDS

This procedure requires that you have available at least a gram of solid sample. The sample should be finely divided, preferably powdered. If it seems advisable, you may grind your sample in a mortar, *testing first* with a tiny portion to make sure that it will not explode.

Step 1. *Solubility characteristics.* Determine the solubility of the sample in the reagents listed, using about 50 mg of solid (enough to cover a $\frac{1}{16}$ inch circle on the end of a small spatula) and about 0.5 ml of solvent in a test tube; test with each of the following reagents:

1. water	3. 6M HNO_3
2. 6M HCl	4. 6M H_2SO_4

With each reagent, stir the solid with your stirring rod for several minutes, noting any reactions that occur, evolution of gases and their odors, and changes in color of the solid or the solution. If the reaction appears to be slow in the cold, put the test tube in a boiling water bath for a maximum of 10 minutes, stirring occasionally. The behavior of the solid in the various solvents may offer several clues to its composition.

In Appendix 1 we have listed the solubility characteristics of most of the salts, hydroxides, and oxides of the cations we have investigated. In particular, we show water solubility, and solubility in strong acids, strong bases, and ammonia. The colors of the cations in solution may be indicative and are listed in Table 5-1. Even if your solid is a mixture, it may be very helpful to compare its behavior with that described in Appendix 1. If you have a strong indication that a given cation or anion is present, it may well be useful to make specific tests for the presence of that ion at this point.

Step 2. *Preparation of a solution for cation analysis.* Since most salts dissolve, at least in principle, in one or more of the solvents used in Step 1, it is very likely that some, or all, of the solid will go into solution in at least one of these solvents. If your solid is soluble in water, you may be able to use the water solution for analysis of all of the cations and some of the anions. If it is soluble in one of the acid reagents, that is probably the best one to use for the cation analysis and some of the anion analysis. If the solid is not completely soluble in any of the reagents, even when hot, select the solvent that appears to be most effective. If there is no apparent difference between the solvent effects of the acids, use 6M HCl.

To a sample of solid weighing about 0.25 g, in a 30 ml beaker, add 2 ml of the solvent that seems most appropriate. If that solvent is water, add 2 ml of 6M HCl in addition to the water. Add 3 ml of water and bring the solvent to a boil; keep it boiling gently until the volume of the liquid decreases to about 2 ml. Then add 4 ml of water, and stir well to dissolve any crystals on the walls of the beaker. Transfer the liquid, plus any undissolved or precipitated solid, to a test tube and centrifuge out any solid. Decant the solution into a test tube; this is the solution we will use for the cation analysis. Wash any undissolved solution twice with 3 ml portions of water. Decant the wash and discard it. Keep the solid for later analysis.

Step 3. Using 0.5 ml portions of the solution obtained in Step 2, carry out the specific anion tests for phosphate and oxalate ions. Before proceeding with these tests, add 6M NaOH, drop by drop, to the solution until it is just basic to litmus. If both phosphate and oxalate ions are absent, you may use the solution from Step 2 for analysis of the Group I to IV cations; if HCl was used in Step 2, you may start the analysis with the Group II cations, because the Group I halides are in the solid that was removed.

Before going through the general analysis, especially if you know your solid has a limited number of components, it is highly advisable to test the solution for the presence of cations that form complex ions with NH_3 or OH^-. This can be done by first adding 2M Na_2CO_3 in excess (blue to litmus) to 0.5 ml of the cation solution. (If there is no precipitate, and the solution is basic to litmus, only Na^+, K^+, and NH_4^+ can be present as cations.) Centrifuge out the carbonate precipitate, discard the liquid, and test the solubility of separate portions of the precipitate in 6M NaOH and in 6M NH_3. Note any observations and compare them with those listed in Appendix 1. Taking account of the results of these tests, perform the necessary parts of the scheme for cation analysis.

Step 4. If either phosphate or oxalate ions are present, they must be separated from the cations or else most or all of the Ba^{2+}, Ca^{2+}, and Mg^{2+} precipitate with the the Group III precipitate in the general analysis. To accomplish this separation, put 3 ml of the cation solution from Step 2 in a 30 ml beaker and add 2M Na_2CO_3 in excess. Bring the solution to a boil and keep it gently boiling for about two minutes. This should precipitate all the heavy metal cations as carbonates, leaving most of the phosphate and oxalate ions in solution. By repeated centrifuging operations, collect all of the solid carbonate precipitate in one test tube, discarding the liquid. Wash the precipitate thoroughly twice with 3 ml portions of water. Centrifuge and discard the wash. Dissolve the solid in 1 ml of 6M HNO_3 and 2 ml of water. Heat the test tube for a few minutes in the boiling water bath to drive out the CO_2 that is liberated. The resulting solution should be free of phosphate and oxalate and can be used for the systematic analysis of the cations present in the solid. Here, as before, it is likely to be helpful to test this solution, or the one from Step 2, for the presence of cations forming amphoteric hydroxides or ammonia complex ions.

Step 5. *Preliminary tests for anions.* Using about 0.1 g of the solid sample, carry out Preliminary Test 1 for the Analysis of Anions. This test may give you strong indications for the presence of CO_3^{2-}, SO_3^{2-}, S^{2-}, NO_2^-, $C_2H_3O_2^-$, and, on heating, ClO_3^-. Record your observations and the anions, if any, that are likely to be in the solid. *If chlorates are absent,* as indicated by the test, repeat the test, using 0.1 g of solid, and a few drops of 18M H_2SO_4. *Be careful,* because this acid is very corrosive when in contact with your clothes and skin. The reagent 18M H_2SO_4 readily reacts with solid halides in the cold, releasing the hydrogen halide as a gas, and in the case of bromides and iodides, forming the elementary halogen. If nitrates are present, they volatilize when the salt-acid mixture is heated, and the HNO_3 evolved partially decomposes, producing the characteristic brown vapors of NO_2. The observations and reactions occurring with 18M H_2SO_4 that are *in addition* to those observed in Preliminary Test 1 for Anions are summarized as follows:

Cl^- Hydrogen chloride is given off, a colorless gas with the odor of 12M HCl. This gas fumes in moist air and turns moist blue litmus red.

Reaction: $MCl(s) + H_2SO_4(l) \longrightarrow HCl(g) + MHSO_4(s)$

Br^- Hydrogen bromide and Br_2 are evolved. Bromine is a reddish brown gas and has a characteristic odor. Hydrogen bromide has properties similar to those of HCl.

Reactions:
$$MBr(s) + H_2SO_4(l) \longrightarrow HBr(g) + MHSO_4(s)$$
$$2\ HBr(g) + H_2SO_4(l) \longrightarrow Br_2(g) + SO_2(g) + 2\ H_2O(l)$$

I^- Hydrogen iodide, I_2, and H_2S are formed. Solid iodine is very dark; the vapor is violet. Hydrogen iodide has properties similar to those of HCl but is much more easily oxidized.

Reactions:
$$MI(s) + H_2SO_4(l) \longrightarrow HI(g) + MHSO_4(s)$$
$$8\ HI(g) + H_2SO_4(l) \longrightarrow 4\ I_2 + H_2S(g) + 4\ H_2O(l)$$

SCN^- A colorless gas, believed to be COS, is evolved in the cold. The liquid may turn yellow.

Reaction: $MSCN(s) + 2\ H_2SO_4(l) + H_2O \longrightarrow$
$$COS(g) + NH_4HSO_4(s) + MHSO_4(s)$$

NO_3^- Nitrogen dioxide, a brown gas similar in color and properties to Br_2, is given off when the mixture is warmed.

Reactions:

$$MNO_3(s) + H_2SO_4(l) \longrightarrow HNO_3(l) + MHSO_4(s)$$
$$4\ HNO_3(l) \longrightarrow 4\ NO_2(g) + O_2(g) + 2\ H_2O(l)$$

$C_2O_4^{2-}$ Carbon monoxide and CO_2 are given off when the mixture is heated. Bubbling the evolved gas through saturated $Ba(OH)_2$ solution precipitates $BaCO_3$.

Reaction: $M_2C_2O_4(s) + H_2SO_4(l) \longrightarrow$
$$M_2SO_4(s) + CO(g) + CO_2(g) + H_2O(l)$$

Of all the anions studied, only sulfate and phosphate do not respond in any way to treatment of the solid with 18M H_2SO_4. If only a few anions are present, the preliminary tests should be very useful in establishing their identity.

Step 6. *Preparation of solutions for anion analysis.* Treat about 0.2 g of the solid sample in a test tube with 2 ml of 6M HCl. Stir well in the cold, for a few minutes at least, to dissolve as much of the solid as possible. Add 4 ml of water and mix thoroughly. Centrifuge out any undissolved solid. Decant the liquid into a test tube. The solid may be discarded. The solution may be used to test for the following anions:

$$SO_4^{2-},\ CrO_4^{2-},\ PO_4^{3-},\ SO_3^{2-},\ S^{2-},\ NO_3^{-},\ and\ NO_2^{-}$$

Keep in mind that the solution is very acidic with HCl, and more HCl does not have to be added in procedures that call for it in the first step. It is possible that some cations may interfere with some of the tests, but that must be accepted and dealt with later. Carbonate should be tested for separately with a small sample of the original solid.

Step 7. Treat about 0.2 g of the solid in a test tube with 4 ml of water and 2 ml of 6M H_2SO_4. Dissolve as much of the solid as possible, stirring well for a few minutes in the cold. Centrifuge out and discard any undissolved solid. The solution may be used to test for any of the following anions:

$$CrO_4^{2-},\ PO_4^{3-},\ S^{2-},\ Cl^{-},\ Br^{-},\ I^{-},\ SCN^{-},\ NO_2^{-},\ NO_3^{-},\ and\ ClO_3^{-}$$

Again, in adapting any procedures to your solution, remember that the solution is strongly acidic with H_2SO_4.

Step 8. *Anion analysis after removal of interfering cations.* Treat 0.3 g of solid with 2 ml of water and 2 ml of the solvent that you used in preparing your solution for cation analysis. Stir for a few minutes to dissolve as much solid as possible. Decant the solution and any solid residue into a 30 ml beaker and add 10 ml of 2M Na_2CO_3, so that carbonate ion is clearly present in excess. Any dissolved cations forming insoluble carbonates precipitate at this point. Boil the slurry gently for about five minutes. By repeatedly centrifuging, remove all the solid precipitate from the system, decanting the clear supernatant liquid into a 30 ml beaker. Most of the anions present in the original solid should not be in this *solution*. Add 6M H_2SO_4 until the solution is acidic, and then boil it gently to drive off all the CO_2, of which there is a great deal. Boil the solution down to a volume of 6 ml. This solution may be used for the detection of the following anions:

$$CrO_4^{2-}, PO_4^{3-}, C_2O_4^{2-}, Cl^-, Br^-, I^-, SCN^-, \text{ and } ClO_3^-$$

Using the results from the Preliminary Tests for Anions, plus those from Steps 6, 7, and 8, it should be possible to establish which anions are present in the original solid, except in special circumstances.

Step 9. *Identification of insoluble residues.* In Step 2 there may have been a residue that did not dissolve in any of the reagents used. This residue is very likely to be one of the following substances:

a. An insoluble sulfate	$BaSO_4$, $PbSO_4$, or possibly $CaSO_4$
b. A silver salt	$AgCl$, $AgBr$, AgI, $AgSCN$
c. A lead salt	$PbSO_4$, $PbCl_2$
d. A mercury salt	Hg_2Cl_2, HgS
e. An oxide	SnO_2, PbO_2, MnO_2, Sb_2O_3
f. An oxychloride	$BiOCl$, $SbOCl$
g. Sulfur	

Assuming, for the moment, that only one substance is present, since that is certainly a likely situation, it is profitable to seek a solvent for that substance. The following reagents may be effective solvents, and have the indicated properties:

6M HCl	dissolves $BiOCl$, $SbOCl$, Sb_2O_3, and, to a slight extent, $AgCl$ and $PbCl_2$
6M H_2SO_4	plus 3 per cent H_2O_2 dissolves MnO_2, PbO_2, plus the oxychlorides
6M NH_3	dissolves $AgCl$
1M $Na_2S_2O_3$	dissolves silver salts
3M $NH_4C_2H_3O_2$	dissolves lead salts, but not PbO_2
aqua regia	dissolves all of the solids except $BaSO_4$, $PbSO_4$, and possibly SnO_2
2M Na_2CO_3	Transposes the sulfates to carbonates

In making the tests, use about 0.5 ml of the reagent. If a reagent is ineffective, even when heated for five minutes in the boiling water bath, centrifuge out the solid, decant the solvent, and wash the solid with 3 ml of water. Centrifuge, and discard the wash. Then proceed with another solvent. In transposing the sulfates, boil them with 3 ml of 2M Na_2CO_3 for five minutes or so. The insoluble carbonate that is formed may be dissolved in 6M HNO_3 and the indicated test performed for positive identification of the cation. If the solid dissolves in a given solvent, carry out specific tests for the cation you believe was present in the insoluble residue. The anion can be tested for, or, in many cases, can be deduced. Do not assume that solubility in a given solvent is a definite test for the given solid.

Step 10. Assemble all your observations and results of analysis and reach a conclusion regarding the composition of the solid sample. Make sure that your sample as you describe it would be expected to have the properties you have observed. You may find it useful at this point, in view of your conclusions, to carry out a few simple specific tests for cations or anions to confirm the results you obtained from the more lengthy procedures.

COMMENTS ON GENERAL PROCEDURE FOR ANALYSIS OF SOLIDS

Although we have listed the steps to follow in numerical order, it is by no means necessary that the steps be carried out in the order given. Your problem is to determine the composition of the solid, and you may find that it is best to first find out which anions are present, rather than which cations, or that some of each are identified before all of either the cations or anions are known. There is probably no one best procedure for analyzing a general solid mixture, and you should use your chemical reasoning as best you can, adapting the suggested procedure to your special case.

Step 1. The general solubility characteristics of a solid may, or may not, be very useful in identifying it. Changes in the color or structure of the solid may indicate that a reaction occurs when a solvent is added. Gases may be evolved that can be identified readily by odor.

It is not necessary that all of the sample be dissolved. Relatively few solids do not go into solution in one of the solvents used, and separate identification of such solids may actually be easier than proceeding to force them into solution and determining their component ions in the larger mixture.

Step 2. The analysis of a solid certainly is most easily accomplished if it is completely soluble in water. If the solution in water is neutral or nearly so, a great many cation-anion combinations can be ruled out and the analysis procedure simplified tremendously. Boiling the acid solution removes many of the anions from the system. Carbonates, sulfites, sulfides, nitrites, and, to some extent, acetates, are volatilized and driven out.

Dilution of the acid with water may cause oxy-salts of antimony and bismuth to precipitate, even if chloride ion is not present.

Step 3. Since Group III cations are precipitated under alkaline conditions, the presence of phosphate or oxalate ions causes barium, calcium, and magnesium to precipitate along with the Group III cations at the point at which Group III cations are separated. This is unfortunate, but it must be recognized. If phosphates and oxalates are absent, the solution from Step 2 may be used for analysis without further treatment.

The determination of the presence or absence of cations forming amphoteric hydroxides or ammonia complex ions is easy and can be very helpful in simplifying the analysis. If the carbonate precipitate is completely soluble in 6M NaOH, the cations that need to be considered are limited to the following: Pb^{2+}, Sn^{2+} or Sn^{4+}, Sb^{3+} or Sb^{5+}, Zn^{2+}, Al^{3+}, and Cr^{3+} (which form complexes with OH^-) plus Na^+, K^+, and NH_4^+ (which do not precipitate with CO_3^{2-}).

If the carbonate precipitate dissolves in 6M NH_3, one can restrict his attention to the following cations: Ag^+, Cu^{2+}, Cd^{2+}, Ni^{2+}, and Zn^{2+} (which form complexes with NH_3) plus Na^+, K^+, and NH_4^+.

Step 4. This step can of course be omitted if PO_4^{3-} and $C_2O_4^{2-}$ are absent. If you are lucky, that will be the case.

Step 5. As you have discovered, Preliminary Test 1 for Anions is quickly done and can afford quick identification of anions that form volatile gases in acid solution. Care in performing this test is worth the effort.

The test with 18M H_2SO_4 must not be done if chlorates are present, because an explosion may result. Otherwise, the test is useful in indicating the presence of halides or, depending on circumstances, nitrates, oxalates, and thiocyanates. Be careful in smelling the vapors, because they are all poisonous, to say the least.

Steps 6 and 7. We use two acid solutions, because analysis for sulfate and sulfite is best carried out in HCl solution and that for the Chloride Group in H_2SO_4 solution. Most of the other anions listed can be determined in either acid. Clearly, if you get a strong positive test for an anion such as CrO_4^{2-} in HCl solution, it is unnecessary to repeat the test in H_2SO_4 solution.

Note that in Steps 6 and 7 the solutions are not heated, to avoid volatilizing anions such as S^{2-} or oxidizing anions such as NO_2^-. This does have the disadvantage that certain salts that would be brought into solution by hot HCl or H_2SO_4 do not dissolve in the cold.

In carrying out some of the tests in these steps, you may find interferences from heavy metal cations. For example, suppose you are attempting to test for SCN^- by adding Fe^{3+} to give a deep red color. If a colored cation, such as Cu^{2+} or Cr^{3+}, is present, your test will be of doubtful value. Again, in order to test for $C_2O_4^{2-}$, it is necessary that the solution be made basic with NH_3 at one point. If ions such as Fe^{3+} or Bi^{3+} are present, you will obtain a heavy precipitate of the corresponding hydroxides, which will be puzzling, to say the least.

You may find that Preliminary Tests 2, 3, 4, and 5 for Anions are useful with the solutions obtained in Steps 6 and 7. It is possible, however, that you will get interferences by cations of the type referred to previously.

Step 8. This step is desirable for unknowns that fail to dissolve in Steps 6 and 7 or give cation interferences as just discussed. The trouble is that the treatment destroys several of the anions by volatilization or oxidation because it involves boiling with acid at one step. Consequently, it has limited applicability.

Step 9. Identification of insoluble solids is relatively easily accomplished, especially if only one solid is present. Once you have found a solvent, the specific tests for the cation are straightforward: Bismuth(III) and Antimony(III) reprecipitate from HCl solution on addition of NH_3 and can then be identified by the standard procedures. Lead(II) reprecipitates from acetate solution on addition of H_2SO_4. Silver(I) precipitates as a black sulfide on heating of the acidified thiosulfate solution. Mercury(II) can be identified by the standard procedure. Barium ion can be reprecipitated by addition of H_2SO_4 to the solution obtained by dissolving $BaCO_3$. Sulfur burns, on the end of a stirring rod, producing SO_2.

PROBLEMS

1. Complete and balance the following equations:
   a. $Al(s) + H^+ \longrightarrow$
   b. $Cu(s) + H^+ + NO_3^- \longrightarrow Cu^{2+} + NO_2(g) + H_2O$
   c. $Au(s) + H^+ + NO_3^- + Cl^- \longrightarrow AuCl_4^- + NO_2(g) + H_2O$
   d. $Ba^{2+} + PO_4^{3-} \longrightarrow$
   e. $I^- + H^+ + SO_4^{2-} \longrightarrow$
   f. $BaSO_4(s) + CO_3^{2-} \longrightarrow$

2. Suggest a reagent that would completely dissolve each of the following mixtures:
   a. $Cu(NO_3)_2$ and KCl d. $AgCl$ and $Zn(OH)_2$
   b. CaO and $Zn(OH)_2$ e. $CaCO_3$ and $NaCl$
   c. Cu and Mg f. $Al(OH)_3$ and Fe

3. Explain why:
   a. In Step 1 of the procedure for dissolving an alloy, the final solution is not tested for Group I cations.
   b. Sodium hydroxide is added in Step 3 of the procedure for dissolving an alloy.
   c. Sodium carbonate is added in Step 3 of the analysis for a general solid unknown.
   d. In preparing solutions for anion analysis (Steps 6 and 7 of the procedures for a general solid unknown) both 6M HCl and 6M H_2SO_4 are used as solvents.

4. State precisely what you would observe in each step of the analysis of an alloy containing only the following:
   a. Ag and Cu   b. Mg and Al

5. State precisely what you would observe in each step of the analysis of a general solid unknown containing only the following:
   a. Na_2SO_4 b. $BiCl_3$ c. ZnO

6. Devise a simplified scheme of analysis for an unknown that may contain only the following solids:
   a. KNO_3 and $BaCO_3$ b. $AgCl$ and $NaOH$ c. CuS, $NaCl$, and CaC_2O_4

7. A solid mixture may contain Cu, ZnO, NaCl, and $BaSO_4$. The solid is partially soluble in water. The residue from treatment with water is completely insoluble in both 6M HCl and 6M HNO_3. Identify the components of the solid.

8. A solid mixture may contain any of the compounds CuS, $PbCl_2$, $Ni(NO_3)_2$, $Fe(OH)_3$, $CaBr_2$, and K_2SO_4. The mixture is white and partially soluble in water. Which solids are definitely present? Which are definitely absent? Which are questionable?

101. Write balanced net ionic equations for the reactions that occur when:
   a. An Al-Zn alloy dissolves in hydrochloric acid (two reactions).
   b. Antimony dissolves in aqua regia to form $SbCl_6^{3-}$ and $NO_2(g)$.
   c. Silver chloride dissolves in 12M HCl.
   d. $CaCl_2(s)$ is treated with $H_2SO_4(l)$.
   e. A solution of sodium carbonate is treated with sulfuric acid.

f. A solution containing Zn^{2+} is treated with CO_3^{2-} to give a precipitate that dissolves in either NaOH or NH_3 (three reactions).

102. Consider the following solids. Which would you expect to dissolve in water? Which would dissolve in 6M HCl but not in water? Which would dissolve in 6M HNO_3 but not in 6M HCl or water?
     a. Mg   b. KBr   c. CuO   d. Ag   e. CuS.

103. Explain why:
     a. In Step 1 of the procedure for dissolving an alloy, the successive solutions are boiled down to small volumes.
     b. It is necessary to remove PO_4^{3-} and $C_2O_4^{2-}$ ions before analyzing for cations in a general solid unknown.
     c. In carrying out the preliminary tests for anions in a general solid unknown, one must test for ClO_3^- before adding H_2SO_4.
     d. Sodium carbonate is used in Step 8 of the procedure for analysis of a general solid unknown.

104. State precisely what you would observe in each step of the analysis of an alloy. containing the following:
     a. Pb and Sn   b. Zn and Ni   c. Bi and Cu

105. State precisely what you would observe in each step of the general procedure for analyzing a solid unknown if it contained the following:
     a. $CuSO_4$ and $Pb(NO_3)_2$ b. AgSCN and $Fe(NO_3)_3$

106. Suggest a simple scheme of analysis for a general solid unknown restricted to the following:
     a. $Pb(NO_3)_2$ and $BaCl_2$ c. AgI, $NaNO_3$ and $Fe(OH)_3$.
     b. $CaCO_3$ and Na_2SO_4

107. A solid mixture may contain $AgNO_3$, $CuSO_4$, KCl, and $Ca(NO_3)_2$. Treatment of the solid with water leaves a white insoluble residue. The water-soluble portion is colorless but gives a precipitate when treated with CO_3^{2-}. Identify the components of the mixture.

108. A solid mixture may contain Al, CaC_2O_4, $PbSO_4$, and $FeCl_3$. The mixture is colored. It is partially soluble in water; the residue dissolves completely in HCl. On the basis of this information, what can you say about the identity of the mixture? What additional tests would you carry out to completely identify the mixture?

LABORATORY ASSIGNMENTS

1. You will be furnished an alloy containing several metallic components. Use the standard procedure to identify any three of the metals present.

2. Determine two of the metals present as major components and two present as minor components in a sample of alloy that will be given you.

3. Determine the cation and anion present in a pure substance to be assigned to you as an unknown.

4. You will be given a solid sample that may contain as many as three cations and three anions. Use any procedure that seems reasonable to determine the cations and anions present in the solid.

5. Your solid unknown will be made up of one to three of the solids listed in the following sets. Use any procedure you wish to determine which solids are contained in your sample. The sets of solids are the following:

   a. $CaCO_3$, Na_2SO_4, CuO, $NH_4C_2H_3O_2$ e. KSCN, $NaHSO_4$, $CrCl_3$, SnO_2
   b. Ag_2SO_4, KNO_3, $NiSO_4$, $Pb(C_2H_3O_2)_2$ f. $BaCl_2$, $Cd(NO_3)_2$, $(NH_4)_2SO_4$
   c. $CoCO_3$, PbO_2, $AlCl_3$, Sb_2O_3 g. MnO_2, $AgNO_3$, $KClO_3$, $FeSO_4$
   d. $BiCl_3$, HgO, K_2CrO_4, CaC_2O_4 h. $NaNO_2$, $MnSO_4$, Hg_2Cl_2, MgO

6. You will be given an unknown that contains four solid ionic substances. Use any procedure you wish to identify the cations and anions present in those solids.

SUMMARY OF SOLUBILITY PROPERTIES OF IONS AND SOLIDS

	Cl^-	SO_4^{2-}	CO_3^{2-}, PO_4^{3-}	CrO_4^{2-}	OH^-	H_2S, $pH = 0.5$	S^{2-}, $pH = 9$
Na^+, K^+, NH_4^+	S	S	S	S	S	S	S
Ba^{2+}	S	I	A	A	S	S	S
Ca^{2+}	S	S^-	A	S	S^-	S	S
Mg^{2+}	S	S	A	S	A	S	S
Fe^{3+}	S	S	A	A	A	S	A
Cr^{3+}	S	S	A, (B)	A, (B)	A, (B)	S	A, (B)
Al^{3+}	S	S	A, B	A, B	A, B	S	A, B
Ni^{2+}	S	S	A, N	A, N	A, N	S	A^+, O^+
Co^{2+}	S	S	A, (N)	A, (N)	A, (N)	S	A^+, O^+
Zn^{2+}	S	S	A, B, N	A, B, N	A, B, N	S	A
Mn^{2+}	S	S	A	(A)	A	S	A
Cu^{2+}	S	S	A, N	A, N	A, N	O	O
Cd^{2+}	S	S	A, N	A, N	A, N	A^+, O	A^+, O
Bi^{3+}	A	A	A	A	A	O	O
Hg^{2+}	S	S	A	A	A	O^+, C	O^+, C
Sn^{2+}, Sn^{4+}	A, B	A, B	A, B	A, B	A, B	A^+, C	A^+, C
Sb^{3+}	A, B	A, B	A, B	A, B	A, B	A^+, C	A^+, C
Ag^+	A^+, N	S^-, N	A, N	A, N	A, N	O	O
Pb^{2+}	HW, B, A^+	B	A, B	B	A, B	O	O
Hg_2^{2+}	O^+	S^-, A	A	(A)	A	O^+	O^+

Key: S, soluble in water.
A, soluble in acid (6M HCl or other nonprecipitating, non-oxidizing acid).
B, soluble in 6M NaOH.
O, soluble in hot 6M HNO_3.
N, soluble in 6M NH_3.
( ), subject to limitations—see section on this ion.

I, insoluble in any common reagent.
S^-, slightly soluble in water.
A^+, soluble in 12M HCl.
O^+, soluble in aqua regia.
C. soluble in 6M NaOH containing excess S^{2-}.
HW, soluble in hot water.

Example: For Cu^{2+} and OH^- the entry is A, N. This means that $Cu(OH)_2(s)$, the product obtained when solutions containing Cu^{2+} and OH^- are mixed, will dissolve to the extent of at least 0.1 mole per liter when treated with 6M HCl or 6M NH_3. Since 6M HNO_3, 12M HCl, and aqua regia are at least as strongly acidic as 6M HCl, $Cu(OH)_2(s)$ would also be soluble in those reagents.

165

COMPLEX ION FORMATION

The following cations form stable complex ions with the indicated ligands:

NH_3	ammonia	Ag^+, Cd^{2+}, Cu^{2+}, Zn^{2+}, Ni^{2+}, (Co^{2+})
OH^-	hydroxide	Pb^{2+}, Sn^{2+}, Sn^{4+}, Sb^{3+}, Zn^{2+}, Al^{3+}, (Cr^{3+})
Cl^-	chloride	(Ag^+), (Pb^{2+}), Hg^{2+}, Cd^{2+}, Sn^{2+}, Sn^{4+}, Sb^{3+}
S^{2-}	sulfide	Hg^{2+}, Sn^{4+}, Sb^{3+}
$S_2O_3^{2-}$	thiosulfate	Ag^+,
$C_2O_4^{2-}$	oxalate	Sn^{4+}, Fe^{3+}

ANSWERS TO PROBLEMS

CHAPTER 1

101. In water solution, $Mg(OH)_2$ exists in equilibrium with its ions: $Mg(OH)_2(s) \rightleftharpoons Mg^{2+} + 2\,OH^-$; $K_{sp} = 1.1 \times 10^{-11} = [Mg^{2+}][OH^-]^2$. In acid solution, $[OH^-] < 1 \times 10^{-7}$ M. To establish equilibrium, $[Mg^{2+}]$ must be at least 1×10^3 M, which is impossible; therefore, $Mg(OH)_2$ must be completely soluble in acid solution.

102. Ammonia reacts with H_2O according to the equation $NH_3 + H_2O \rightleftharpoons NH_4^+ + OH^-$; $K_b = 2 \times 10^{-5} = [NH_4^+][OH^-]/[NH_3]$. In 1 lit of 0.1M NH_3 there is initially 0.1 mole of NH_3; x moles react to form x moles of NH_4^+ and x moles of OH^-. Since K_b is small, x is likely to be negligible compared to 0.1. In the equilibrium condition: $[NH_4^+][OH^-]/[NH_3] = (x)^2/0.1 = 2 \times 10^{-5}$; $x = 1.4 \times 10^{-3} = [OH^-] = [NH_4^+]$. Since about 1.4 per cent of the NH_3 is ionized, x indeed is much less than 0.1.

103. In the buffer initially $[NH_3] = 1M$ and $[NH_4^+] = 1M$. These species are in equilibrium: $NH_3 + H_2O \rightleftharpoons NH_4^+ + OH^-$; $K_b = 2 \times 10^{-5}$. About 2×10^{-5} moles of NH_3 react with H_2O to form OH^- and NH_4^+ to establish equilibrium in the initial buffer. When 0.1 mole of OH^- is added to the buffer, it reacts with NH_4^+: $NH_4^+ + OH^- \longrightarrow NH_3 + H_2O$; $K = 5 \times 10^4$. This adds 0.1 mole of NH_3 to, and removes 0.1 mole of NH_4^+ from, the solution; at equilibrium, $[NH_3] = 1.1M$ and $[NH_4^+] = 0.9M$. The $[OH^-]$ must therefore be $(2 \times 10^{-5})(1.1/0.9) = 2.2 \times 10^{-5}$ M, only a 10 per cent change.

104. For calcium carbonate: $CaCO_3(s) + 2\,H^+ \rightleftharpoons Ca^{2+} + CO_2 + H_2O$; $K = K_{sp}/K_1K_2 = 2.5 \times 10^8$. The K for this reaction is large; reaction goes to completion, using up just about all the H^+ from the acid, namely, 0.2 mole. In this reaction, 0.1 mole of $CaCO_3$ must dissolve to react with the H^+.

105. Silver carbonate reacts with H^+: $Ag_2CO_3(s) + 2\,H^+ \longrightarrow 2\,Ag^+ + CO_2 + H_2O$; $K = 3 \times 10^5$. There is no analogous reaction for AgCl, since HCl is a strong acid; for the reaction of AgCl with H^+, K is essentially zero.

167

106. For silver chloride: $AgCl(s) + 2 NH_3 \rightleftharpoons Ag(NH_3)_2^+ + Cl^-;$ $K = [Ag(NH_3)_2]^+[Cl^-]/[NH_3]^2 = 5 \times 10^{-3}$. If we dissolve 0.1 mole of $AgCl$ in 1 lit of solution, at equilibrium $[Cl^-] = [Ag(NH_3)_2]^+ = 0.1M;$ $[NH_3]^2 = 0.1 \times 0.1/(5 \times 10^{-3}) = 2;$ $[NH_3] = 1.4M$ at equilibrium. Since 0.2 mole of NH_3 was used up in forming the complex, the original concentration of NH_3 had to be at least 1.6M.

CHAPTER 3

GROUP I CATIONS

101. a. $Hg_2^{2+} + 2 Cl^- \longrightarrow Hg_2Cl_2(s)$
    b. $Ag(NH_3)_2^+ + 2 H^+ + Br^- \longrightarrow AgBr(s) + 2 NH_4^+$
    c. $2 Ag^+ + 2 OH^- \longrightarrow Ag_2O(s) + H_2O$
    d. $Pb^{2+} + 2 Cl^- \longrightarrow PbCl_2(s)$
    e. $Ag^+ + Cl^- \longrightarrow AgCl(s)$
       $AgCl(s) + Cl^- \longrightarrow AgCl_2^-$

102. a. $3 Ag(s) + NO_3^- + 4 H^+ \longrightarrow 3 Ag^+ + NO(g) + 2 H_2O$
    b. $Hg_2^{2+} + H_2S \longrightarrow Hg(l) + HgS(s) + 2 H^+$
    c. $3 Pb^{2+} + ClO_3^- + 6 OH^- \longrightarrow 3 PbO_2(s) + Cl^- + 3 H_2O$

103. a. Water.   c. Nitric acid.
    b. Concentrated ammonia.   d. Boil with sodium carbonate and
                                        acidify.

104. a. Add NH_3; only AgCl dissolves.
    b. Add Cl^-; a precipitate forms if $Hg_2(NO_3)_2$ is present.
    c. Heat with water and test the solution with CrO_4^{2-}; the precipitate is $PbCrO_4$.
    d. Add water; only $Pb(NO_3)_2$ dissolves.
    e. Add HCl; then heat the precipitate with water and test for Pb^{2+}. Alternatively, you could add NH_3 and test the solution for Ag^+.
    f. Add 0.1M NaCl; only Ag^+ gives a precipitate.
    g. Add OH^- (small amount); Pb^{2+} precipitates.
    h. Add Cl^-; only Hg_2^{2+} precipitates.

105. a. It gives no information except that at least one Group I ion is present.
    b. Silver is present.
    c. These observations are inconsistent; they imply that the Pb^{2+} was lost between the two steps or that the first observation was in error.
    d. The observations are inconsistent; apparently the cation that gave the original precipitate has been lost in the analysis.

106. Step 1: Precipitate forms.
     Step 2: Part of precipitate dis-
        solves.

     Step 3: Yellow precipitate appears.
     Step 4: Precipitate turns black.
     Steps 5 and 6: Negative test results.

107. a. The test for Pb^{2+} may be weak; the precipitate is heavily contami-
        nated with $PbCl_2$.
     b. Any $PbCl_2$ remaining may be confused with Hg_2Cl_2.
     c. Precipitate of AgCl will not form.

108. a. (1) Add HCl to get a precipi-
            tate.
         (2) Heat with water.
     b. (1) Add HCl to get a precipi-
            tate.
         (2) Heat with water.
         (3) Add CrO_4^{2-} to test for
            Pb^{2+}.
         (4) Add NH_3.

     (3) Add CrO_4^{2-} to test for Pb^{2+}.
     (4) Add NH_3; if a precipitate is ob-
        tained, Hg_2^{2+} must be present.
     (5) Add HNO_3 to reprecipitate AgCl.
        If time is of the essence, the fourth
        and fifth steps might be omitted;
        any precipitate remaining after
        the second step should be AgCl.
        Similarly, in part (a), one could
        stop after the third step; any pre-
        cipitate remaining should be
        Hg_2Cl_2.

109. The Hg_2^{2+} ion is present; Pb^{2+} is absent; Ag^+ is questionable.

110. $PbCl_2(s)$, $Hg_2(NO_3)_2(s)$, $Ag_2S(s)$, $PbCO_3(s)$

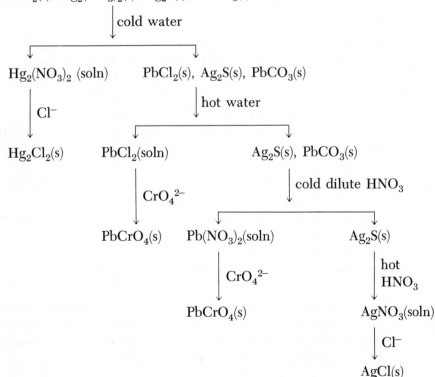

GROUP II CATIONS

101. a. $2 Cu^{2+} + Fe(CN)_6^{4-} \longrightarrow Cu_2Fe(CN)_6(s)$
b. $Bi^{3+} + H_2O + Cl^- \longrightarrow BiOCl(s) + 2 H^+$
c. $Cd^{2+} + 2 OH^- \longrightarrow Cd(OH)_2(s)$
   $Cd(OH)_2(s) + 4 NH_3 \longrightarrow Cd(NH_3)_4^{2+} + 2 OH^-$
d. $Hg^{2+} + 2 I^- \longrightarrow HgI_2(s)$
   $HgI_2(s) + 2 I^- \longrightarrow HgI_4^{2-}$
e. $SnS_2 + 6 OH^- \longrightarrow Sn(OH)_6^{2-} + 2 S^{2-}$

102. a. $2 Cu^{2+} + 4 I^- \longrightarrow 2 CuI(s) + I_2$
b. $3 CdS(s) + 8 H^+ + 2 NO_3^- \longrightarrow 3 Cd^{2+} + 3 S(s) + 2 NO(g) + 4 H_2O$
c. $2 Hg^{2+} + Sn^{2+} + 2 Cl^- \longrightarrow Hg_2Cl_2(s) + Sn^{4+}$
d. $Sn(s) + 4 NO_3^- + 4 H^+ \longrightarrow SnO_2(s) + 4 NO_2(g) + 2 H_2O$

103. a. None   b. All

104. a. $3 PbS(s) + 8 H^+ + 2 NO_3^- \longrightarrow 3 Pb^{2+} + 3 S(s) + 2 NO(g) + 4 H_2O$
b. $Bi_2S_3(s) + 8 H^+ + 2 NO_3^- \longrightarrow 2 Bi^{3+} + 3 S(s) + 2 NO(g) + 4 H_2O$
c. $3 HgS(s) + 12 Cl^- + 8 H^+ + 2 NO_3^- \longrightarrow 3 HgCl_4^{2-} + 3 S(s) +$
   $2 NO(g) + 4 H_2O$
d. $Sb_2S_3(s) + 8 OH^- \longrightarrow 2 Sb(OH)_4^- + 3 S^{2-}$

105. a. Dissolve in HNO_3; add NH_3 to get a blue color with Cu^{2+} or a white
     precipitate with Bi^{3+}.
b. The $Cd(OH)_2$ dissolves in NH_3 or dilute HCl.
c. Silver chloride dissolves in NH_3.
d. Color (HgI_2 is red) or solubility (HgI_2 is insoluble in water).
e. Test the reducing ability (Sn^{2+} reduces Hg^{2+} to Hg_2^{2+} or Hg).
f. Add HCl or H_2SO_4 to give a precipitate with Pb^{2+}.
g. Add Cu^{2+} to give a precipitate with S^{2-}.
h. Add S^{2-} to give a black precipitate (Bi^{3+}) or an orange precipitate
   (Sb^{3+}).

106. a. Heat with HNO_3.
b. Treat it with Sn^{2+} in basic solution.
c. Dissolve it in water, add $Cu(NO_3)_2$, filter off the $Cu(OH)_2$, and evaporate the solution.
d. Use an oxidizing agent such as Hg^{2+}.
e. Use a reducing agent such as Sn(s), Fe(s), or Al(s).

107. a. Step 2: A dark precipitate   Step 12: A precipitate is obtained.
            appears.   Step 13: Part of the precipitate
         Step 3: The precipitate com-   dissolves; the residue is
            pletely dissolves.   black.

Step 14: Black residue dissolves.

Step 15: Positive test reaction for Hg^{2+}.

b. Step 2: A black precipitate appears.

Step 3: Part of the precipitate dissolves.

Step 4: The remainder of the precipitate dissolves.

Step 5: A blue solution and a white precipitate appear.

Step 6: There is a white precipitate of $PbCl_2$.

Step 17: A gray precipitate forms.

Step 18: An orange precipitate appears.

Step 8: White precipitate is present.

Step 9: No precipitate is present.

Step 10: Red-brown precipitate appears.

Step 12: Brownish precipitate is present.

Step 13: The precipitate dissolves.

Step 17: White precipitate is present.

Step 18: No precipitate is present.

108. a. Group III sulfides, being more soluble, fail to precipitate at low S^{2-} concentration.
   b. The Sn^{4+} ion forms hydroxide and sulfide complexes more readily than does the Cu^{2+} ion.
   c. The ion could be Pb^{2+}.
   d. The Pb^{2+} ion forms a soluble complex with Ac^-.
   e. The Sn^{4+} ion is complexed by the oxalate ion; it fails to precipitate with S^{2-}.

109. a. It could fail to precipitate some of Group II sulfides, particularly CdS.
   b. It would not separate Sn^{4+}, Sb^{5+}, and Hg^{2+} from other ions.
   c. It would fail to precipitate Cd^{2+} as $Cd(OH)_2$.
   d. Mercuric sulfide will not go into solution and, hence, no test for Hg(II) will be obtained.

110. a. Add HNO_3; precipitate with H_2S in acidic solution. Dissolve sulfide in NaOH and then follow steps 13 to 18 of the Group II procedure.
   b. Add HNO_3; precipitate with H_2S in acidic solution. Heat with NaOH to dissolve sulfides of Sn, and Sb. Dissolve the remaining sulfide precipitate in HNO_3; add NH_3 to give $Bi(OH)_3$. Reprecipitate sulfides of Sn and Sb by adding HCl, and then heat with concentrated HCl to redissolve. Add Al wire to reduce Sn^{4+} to Sn^{2+}; test for the latter with $HgCl_2$. To another portion of solution add oxalic acid to complex tin and reprecipitate Sb_2S_3.

111. a. Add HCl to precipitate Hg_2Cl_2. To the solution add HNO_3 and H_2S to precipitate sulfides in acidic solution. Heat with NaOH to dissolve

SnS_2. Dissolve residue by heating with HNO_3; add NaOH to give a white precipitate of $Cd(OH)_2$. Reprecipitate SnS_2 by adding HCl to the basic solution.

b. Add HCl to precipitate AgCl. To the solution add HNO_3 and H_2S to precipitate sulfides in acidic solution. Heat with NaOH to dissolve SnS_2 and Sb_2S_3. Dissolve the residue by heating with HNO_3; add NH_3 to give a blue solution if Cu^{2+} is present. Reprecipitate sulfides of Sn and Sb by adding HCl; then heat with concentrated HCl to redissolve. Carry out steps 17 and 18 to test for Sn^{2+} and Sb^{3+}.

112. Absent: Cu^{2+}, Pb^{2+}, Bi^{3+}, Hg^{2+}
     Questionable: Sb^{3+}, Sn^{2+}
     Present: Cd^{2+}

113. Absent: Cu^{2+}, Pb^{2+}, Bi^{3+}, Sb^{3+}, Sn^{2+}
     Present: Cd^{2+}, Hg^{2+}

114. Treat with water; any black residue is CuS. A white residue indicates that $AgNO_3$ and $SnCl_4$ are present; if it does not form, either or both of them must be absent.

    Add HCl to precipitate any Ag^+ present. Add HNO_3 and then H_2S to precipitate SnS_2 and CdS. Heat the sulfide with NaOH; acidify the solution formed to reprecipitate tin as the sulfide. The presence of a residue that is insoluble in NaOH implies that Cd^{2+} is present; confirm if desired.

GROUP III CATIONS

101. a. $Fe^{2+} + H_2S \longrightarrow FeS(s) + 2\,H^+$
    b. $Fe^{3+} + SCN^- \longrightarrow FeSCN^{2+}$
    c. $Al^{3+} + 3\,OH^- \longrightarrow Al(OH)_3(s)$
    d. $Cu^{2+} + H_2S \longrightarrow CuS(s) + 2\,H^+$
    e. $Zn(OH)_2(s) + 4\,NH_3 \longrightarrow Zn(NH_3)_4{}^{2+} + 2\,OH^-$
    f. $2\,BaCrO_4(s) + 2\,H^+ \longrightarrow 2\,Ba^{2+} + Cr_2O_7{}^{2-} + H_2O$
    g. $Ni^{2+} + 2\,OH^- \longrightarrow Ni(OH)_2(s)$
       $Ni(OH)_2(s) + 6\,NH_3 \longrightarrow Ni(NH_3)_6{}^{2+} + 2\,OH^-$

102. a. $4\,Fe(OH)_2(s) + O_2(g) + 2\,H_2O \longrightarrow 4\,Fe(OH)_3(s)$
    b. $2\,Fe^{3+} + 2\,I^- \longrightarrow 2\,Fe^{2+} + I_2$
    c. $Zn(s) + 2\,H^+ \longrightarrow Zn^{2+} + H_2(g)$
    d. $Mn(OH)_2(s) + Br_2 + 2\,OH^- \longrightarrow MnO_2(s) + 2\,Br^- + 2\,H_2O$
    e. $2\,Cr^{3+} + 3\,H_2O_2 + 10\,OH^- \longrightarrow 2\,CrO_4{}^{2-} + 8\,H_2O$
    f. $4\,Co^{2+} + O_2(g) + 24\,NH_3 + 2\,H_2O \longrightarrow 4\,Co(NH_3)_6{}^{3+} + 4\,OH^-$

103. a. O.A, R.A. f. O.A.
 b. O.A. g. O.A.
 c. O.A, R.A. h. R.A.
 d. O.A., R.A. i. O.A.
 e. O.A.

104. NH_3: $Cu(OH)_2$, $Cd(OH)_2$, $Ni(OH)_2$, $Zn(OH)_2$.
 NaOH: $Sn(OH)_2$, $Sn(OH)_4$, $Sb(OH)_3$, $Sb(OH)_5$, $Cr(OH)_3$, $Al(OH)_3$, $Zn(OH)_2$, $Pb(OH)_2$.

105. a. Add I^-; I_2 is formed by Fe^{3+}.
 b. Add excess NH_3; Cu^{2+} gives a deep blue color.
 c. Evaporate to dryness; $Zn(NH_3)_4^{2+}$ gives off NH_3.
 d. Color (blue vs. red).
 e. Add excess OH^-; $Al(OH)_3$ dissolves.
 f. Ferrous sulfide dissolves in HCl.
 g. The compound SnS dissolves in NaOH.
 h. Add Cl^- to precipitate AgCl.

106. a. Add Zn to HCl; evaporate the solution.
 b. Oxidize with air or H_2O_2 in the presence of a complexing agent, such as NH_3.
 c. Treat with H_2S or NaI.
 d. Treat with $NaBiO_3$.
 e. Add acid and then excess NH_3.
 f. Allow it to stand in air.
 g. Oxidize with H_2O_2 in basic solution; acidify to convert CrO_4^{2-} to $Cr_2O_7^{2-}$.
 h. Treat with CN^-.

107. a. Step 1: A dark precipitate appears.
 Step 2: The precipitate partially dissolves, leaving a black residue.
 Step 3: Nickel sulfide dissolves.
 Step 4: The test result is negative.
 Step 5: A red color is present.
 Step 6: There is a yellow solution and a red precipitate.
 Step 7: The precipitate dissolves.
 Step 10: A red color is present.
 Step 12: No precipitate is present.
 Step 13: Results are negative.
 Step 14: A yellow precipitate is present.
 Step 15: Test results are positive.

 b. Step 1: A dark precipitate forms.
 Step 2: The precipitate partially dissolves.
 Step 3: The compound CoS dissolves.
 Step 4: The test results are positive.

Step 5: The test results are negative.

Step 6: A black precipitate and a colorless solution are present.

Step 7: The precipitate remains.

Step 8: The precipitate dissolves.

Step 9: A purple color appears.

Step 10: No color is present.

Step 12: A white precipitate appears.

Step 13: Test results are positive.

108. a. In step 12, $Al(OH)_3$ is not detected. (The pH must be carefully controlled).

b. The Cr^{3+} ion is confused with the Al^{3+} ion.

109. a. Oxidize with H_2O_2 in basic solution to get $Fe(OH)_3$, $Ni(OH)_2$, and $Al(OH)_4^-$. Acidify the solution; then add NH_3 to precipitate $Al(OH)_3$. Add NH_3 to the precipitate of $Fe(OH)_3$ and $Ni(OH)_2$. A red precipitate that fails to dissolve indicates Fe^{3+}; a blue solution indicates Ni^{2+}.

b. Add H_2O_2 in basic solution to get $Fe(OH)_3$, MnO_2, and $Zn(OH)_4^{2-}$. Addition of acid and then $K_4Fe(CN)_6$ to the solution results in a precipitate if Zn^{2+} is present. To the precipitate of $Fe(OH)_3$ and MnO_2, add H_2SO_4 to bring Fe^{3+} into solution; add SCN^- to give a red color. Any precipitate remaining after treatment with H_2SO_4 indicates Mn.

110. a. Add HCl; if Ag^+ is present, AgCl forms. Saturate the HCl solution with H_2S to precipitate CdS and CuS. Dissolve this precipitate in HNO_3; then add NH_3, which gives a blue color if Cu^{2+} is present. Add NaOH to ammoniacal solution; a white precipitate forms if Cd^{2+} is present. To detect Cr^{3+}, add dilute NaOH to the solution remaining after precipitation of Cd^{2+} and Cu^{2+}; a green precipitate indicates Cr^{3+}.

b. Add HCl to precipitate $PbCl_2$. Saturate HCl solution with H_2S to precipitate Sb_2S_3 and PbS. Dissolve the precipitate in NaOH; acidify it again to precipitate Sb_2S_3 if Sb^{3+} is present. To the $HCl-H_2S$ solution, add NH_3 and NH_4Cl to precipitate CoS and ZnS. Add HCl to the sulfide precipitate; a black insoluble residue indicates Co^{2+}. Test the solution for Zn^{2+} as in the sixteenth step of Group III analysis.

111. Present: Fe^{3+}.
Absent: Co^{2+}, Ni^{2+}, Cr^{3+}, Mn^{2+}.
Questionable: Zn^{2+}, Al^{3+}.

112. a. Work with the solution from H_2O_2—NaOH treatment. Acidify; add NH_3 to precipitate $Al(OH)_3$, if present. Test the NH_3 solution as in Step 16 for Zn^{2+}.

b. Dissolve the precipitate from H_2O_2—NaOH treatment in acid; add SCN^-.

113. Present: Ag^+, Co^{2+}, Fe^{3+}.
     Absent: Pb^{2+}, Hg_2^{2+}, all Group II ions, Ni^{2+}, Mn^{2+}.
     Questionable: Cr^{3+}, Al^{3+}, Zn^{2+}.

114. Ag^+: $Ag^+ + Cl^- \longrightarrow AgCl(s)$
     $AgCl(s) + 2\,NH_3 \longrightarrow Ag(NH_3)_2^+ + Cl^-$
     Co^{2+}: $Co^{2+} + H_2S \longrightarrow CoS(s) + 2\,H^+$
     Fe^{3+}: $2\,Fe^{3+} + 3\,H_2S \longrightarrow 2\,FeS(s) + S(s) + 6\,H^+$
     $FeS(s) + 2\,H^+ \longrightarrow Fe^{2+} + H_2S$
     $2\,Fe^{2+} + H_2O_2 + 4\,OH^- \longrightarrow 2\,Fe(OH)_3(s)$

115. Present: Cu^{2+}, Pb^{2+}, Ni^{2+}.
     Absent: Al^{3+}.

116. Absent: Al_2S_3, PbS, Bi_2S_3, NiS.
     Present: FeS.

GROUP IV CATIONS

101. a. No reaction.
     b. $BaCO_3(s) + 2\,H^+ \longrightarrow Ba^{2+} + CO_2(g) + H_2O$
     c. $BaHPO_4(s) + H^+ \longrightarrow Ba^{2+} + H_2PO_4^-$
     d. No reaction.
     e. $Mg^{2+} + 2\,OH^- \longrightarrow Mg(OH)_2(s)$
     f. $NH_3 + H^+ \longrightarrow NH_4^+$

102. a. Add SO_4^{2-} to precipitate $BaSO_4$.
     b. Add Na_2CO_3 to precipitate $CaCO_3$.
     c. Add SO_4^{2-} to precipitate $BaSO_4$.
     d. Add OH^- to precipitate $Mg(OH)_2$.
     e. Flame test for K^+.
     f. Add acid to dissolve $BaCO_3$.
     g. Color.
     h. Add OH^- and heat; smell NH_3.
     i. Test pH; NH_3 is basic, and NH_4^+ is acidic.
     j. Add Ag^+; KCl precipitates.

103. a. Add Cl^- to precipitate AgCl.
     b. Add S^{2-} to precipitate Sb_2S_3.
     c. Add S^{2-} to precipitate CoS.
     d. Add Ba^{2+} to precipitate $BaSO_4$.
     e. Test water solubility.
     f. Add S^{2-} to precipitate Bi_2S_3.
     g. Heat strongly; $Ca(OH)_2$ gives off water.
     h. Add H^+; $BaCrO_4$ dissolves.

104. a. Dissolve in HCl; evaporate.
     b. Dissolve in HCl; evaporate.
     c. Heat with 2M Na_2CO_3; extract Na_2SO_4 with water.

d. Add an equivalent amount of $Ba(NO_3)_2$; filter off $BaCrO_4$; evaporate.
e. Add excess HCl; evaporate.
f. Add HCl; heat.

105. a. HCl, HNO_3.   d. NH_3, NaOH.
     b. HCl, H_2SO_4.   e. HCl, HNO_3.
     c. HCl, NH_3.

106. a. Extract with water; $CaCl_2$ is soluble.
     b. Extract with water; $Ba(OH)_2$ is soluble.
     c. Treat with NH_3; $Cu(OH)_2$ is soluble.
     d. Treat with NaOH: SnS is soluble.

107. a. Reagents used to remove other ions contain NH_4^+ and Na^+.
     b. Ammonia may be driven off, making the pH too low for all ions to precipitate.
     c. The $BaCrO_4$ would not precipitate in the presence of a strong acid.
     d. This converts $Cr_2O_7^{2-}$ to CrO_4^{2-}.
     e. The flame test for Na^+ is very sensitive; trace impurities may give positive test.

108. a. Step 2: A white precipitate is present.
        Step 3: There is a yellow precipitate.
        Step 4: No precipitate is present.
        Step 5: There is a white precipitate and a positive blue lake test.
        Step 6: Tests are positive.
        Step 7: The flame is violet.
        Step 8: No NH_3 is evolved.

     b. Step 2: A white precipitate is present.
        Step 3: No precipitate is present.
        Step 4: No precipitate is present.
        Step 5: There is a white precipitate and a positive blue lake test.
        Step 6: Tests are positive.
        Step 7: There is no flame test.
        Step 8: Ammonia is evolved.

109. a. Add NH_3; if a precipitate is obtained, Mg^{2+} must be present. Add $C_2O_4^{2-}$ to the solution to precipitate Ca^{2+}.
     b. Add SO_4^{2-} to precipitate Ba^{2+}, if present. Flame test original solution for Na^+. Boil the original solution with OH^- to test for NH_4^+.
     c. Flame test for K^+. Make the unknown strongly basic; heat; test for NH_3.

110. a. Add HCl to precipitate Pb^{2+}. Add H_2SO_4 to the solution to precipitate Ba^{2+}. To the remaining solution add excess NH_3 to precipitate $Fe(OH)_3$. The solution will be deeply colored if Cu^{2+} is present, because of $Cu(NH_3)_4^{2+}$; $Zn(NH_3)_4^{2+}$ (colorless) will also be in the solu-

tion if Zn^{2+} is present. Acidify the solution to destroy the complexes. Add excess NaOH to remove $Cu(OH)_2$. Zinc is now in the form of $Zn(OH)_4{}^{2-}$; add acid and then $K_4Fe(CN)_6$ to give a white precipitate if Zn^{2+} is present.

111. Present: Ca^{2+}, K^+.
Absent: Ba^{2+}, Na^+.
Questionable: Mg^{2+}, $NH_4{}^+$.

112. Present: Bi^{3+}, Ca^{2+}.
Absent: Sn^{2+}, Cr^{3+}.
Questionable: $NH_4{}^+$.

113. Present: $NH_4{}^+$, K^+, Sn^{2+}, Ba^{2+}.
Absent: Ag^+, Pb^{2+}, $Hg_2{}^{2+}$, Cu^{2+}, Bi^{3+}, Cd^{2+}, Hg^{2+}, Fe^{3+}, Ni^{2+}, Co^{2+}, Cr^{3+}, Mn^{2+}.
Questionable: Na^+, Sb^{3+}, Al^{3+}, Zn^{2+}.

114. Extract with cold water to remove $BaCl_2$; test the water solution with $CrO_4{}^{2-}$ to see if Ba^{2+} is present. Heat the remaining solid with water to remove $PbCl_2$; test the hot solution with $CrO_4{}^{2-}$ to detect Pb^{2+}. Add NaOH to the solid remaining to bring $Al(OH)_3$ into solution; test for Al^{3+} by acidifying the solution and then slowly adding NH_3 to precipitate $Al(OH)_3$.

    Any solid remaining after treatment with NaOH must be $Fe(OH)_3$ or $Mg(OH)_2$. The color is indicative of $Fe(OH)_3$. Dissolve the solid in HCl. Add NH_3 until the solution is slightly basic; $Fe(OH)_3$ but not $Mg(OH)_2$ will precipitate. Add NaOH to precipitate $Mg(OH)_2$.

CHAPTER 4

PRELIMINARY TESTS FOR ANIONS

101. a. $SO_3{}^{2-} + 2\,H^+ \longrightarrow SO_2(g) + H_2O$
b. $C_2H_3O_2{}^- + H^+ \longrightarrow HC_2H_3O_2$
c. $Ba^{2+} + CrO_4{}^{2-} \longrightarrow BaCrO_4(s)$
d. $Ba^{2+} + SO_3{}^{2-} \longrightarrow BaSO_3(s)$
e. $Ag^+ + SCN^- \longrightarrow AgSCN(s)$

102. a. $6\,I^- + 2\,CrO_4{}^{2-} + 16\,H^+ \longrightarrow 3\,I_2 + 2\,Cr^{3+} + 8\,H_2O$
b. $6\,I^- + ClO_3{}^- + 6\,H^+ \longrightarrow 3\,I_2 + Cl^- + 3\,H_2O$
c. $2\,MnO_4{}^- + 10\,Br^- + 16\,H^+ \longrightarrow 2\,Mn^{2+} + 5\,Br_2 + 8\,H_2O$
d. $2\,MnO_4{}^- + 10\,I^- + 16\,H^+ \longrightarrow 2\,Mn^{2+} + 5\,I_2 + 8\,H_2O$
e. $2\,MnO_4{}^- + 5\,NO_2{}^- + 6\,H^+ \longrightarrow 2\,Mn^{2+} + 5\,NO_3{}^- + 3\,H_2O$

103. a. Add acid; H_2S has a foul odor.
 b. Add $BaCl_2$ to give a precipitate with SO_3^{2-}.
 c. Add $KMnO_4$; it is decolorized by NO_2^-.
 d. Add I^-; it is oxidized by ClO_3^-.
 e. Add Ag^+.
 f. Add $KMnO_4$; it is decolorized by $C_2O_4^{2-}$.
 g. Add H^+ to get H_2S.
 h. Add Ba^{2+} to get $BaSO_4$.
 i. Add $KMnO_4$; it is decolorized by SO_3^{2-}.
 j. Add Cu^{2+}; it gives a precipitate with I^-.

104. a. Oxidize it with $KMnO_4$.
 b. Add acid.
 c. Reduce it with I^-.
 d. Oxidize it with $KMnO_4$.

105. b, d, f, g.

106. a. The MnO_4^- ion has a characteristic color, which fades on reduction.
 b. The I^- is oxidized to I_2, which has a characteristic color; the rate of oxidation also varies with the oxidizing species.
 c. It is not likely unless a Group I cation is present.
 d. Destroy S^{2-}, which would interfere with the test.

107. a. Step 1: There is an odor of vinegar.
 Step 2: Free I_2 is formed.
 Step 3: Results are negative.
 Step 4: A yellow precipitate is present.
 Step 5: There is a white precipitate.
 b. Step 1: There is a foul odor.
 Step 2: Results are negative.
 Step 3: The purple color fades.
 Step 4: Results are negative.
 Step 5: A white (pale yellow) precipitate forms.

108. a. The CO_3^{2-}, SO_3^{2-}, and S^{2-} ions are absent.
 b. The CrO_4^{2-} or NO_2^- ion is present.
 c. The S^- and possibly other oxidizable anions are present.
 d. The I^- ion is present.

109. Present: NO_3^-, SO_4^{2-}.
 Absent: $C_2H_3O_2^-$, I^-.

110. Add acid; if a gas is evolved, CO_3^{2-} is present; if an odor is detected, $C_2H_3O_2^-$ is present.
 Add $BaCl_2$ to give a precipitate with CrO_4^{2-}; Co_3^2 interferes.
 Add $AgNO_3$ in acidic solution to precipitate Cl^-.

If CrO_4^{2-} is present, remove it by adding Ba^{2+}; then test the solution for NO_3^- with I^-.

SULFATE GROUP ANIONS

101. a. $CO_2(g) + Ca^{2+} + 2\,OH^- \longrightarrow CaCO_3(s) + H_2O$
    b. $Pb^{2+} + SO_4^{2-} \longrightarrow PbSO_4(s)$
    c. $2\,BaCrO_4(s) + 2\,H^+ \longrightarrow 2\,Ba^{2+} + Cr_2O_7^{2-} + H_2O$
    d. $BaHPO_4(s) + H^+ \longrightarrow Ba^{2+} + H_2PO_4^-$
    e. $SO_3^{2-} + 2\,H^+ \longrightarrow SO_2(g) + H_2O$
    f. $2\,Ag^+ + SO_3^{2-} \longrightarrow Ag_2SO_3(s)$
    g. $Ba^{2+} + C_2O_4^{2-} \longrightarrow BaC_2O_4(s)$

102. a. $Cr_2O_7^{2-} + 6\,Br^- + 14\,H^+ \longrightarrow 2\,Cr^{3+} + 3\,Br_2 + 7\,H_2O$
    b. $2\,MnO_4^- + 5\,SO_3^{2-} + 6\,H^+ \longrightarrow 2\,Mn^{2+} + 5\,SO_4^{2-} + 3\,H_2O$
    c. $Ag_2S_2O_3(s) + H_2O \longrightarrow Ag_2S(s) + SO_4^{2-} + 2\,H^+$

103. a. Add acid; CO_3^{2-} evolves CO_2.
    b. Add SO_4^{2-}; Ba^{2+} precipitates as $BaSO_4$.
    c. Add Ba^{2+} in strong acid; SO_4^{2-} precipitates.
    d. Add H_2O; Na_2CO_3 is very basic.
    e. Add acid; Na_2SO_3 evolves SO_2.
    f. Add water; $NaHSO_3$ is acidic to litmus.

104. a. HCl.   c. HCl.
    b. NaOH.   d. NH_3.

105. a. The CO_3^{2-} ion is formed from the CO_2 of air in basic solution.
    b. The solution is strongly acidic.
    c. It oxidizes SO_3^{2-} to SO_4^{2-} and precipitates Ba^{2+} as $BaSO_4$.

106. a. One would not obtain a precipitate (Na_2CO_3 is soluble).
    b. One would not get CrO_5.
    c. It would precipitate other anions, such as CO_3^{2-}.

107. a. Add Ba^{2+} in strongly acidic solution to test for SO_4^{2-}. Add HNO_3 and H_2O_2 to obtain CrO_5 as the test for CrO_4^{2-}. Add HNO_3 and $(NH_4)_2MoO_4$ to test for PO_4^{3-}.
    b. Add HCl to give the odor of SO_2 if SO_3^{2-} is present. Add HCl and H_2O_2 to give CO_2. (Test by passing it into $Ba(OH)_2$). Add HNO_3 and H_2O_2 to get a blue color if CrO_4^{2-} is present.

108. Present: CrO_4^{2-}, SO_3^{2-}.
    Absent: SO_4^{2-}.
    Questionable: CO_3^{2-}, PO_4^{3-}, $C_2O_4^{2-}$.

109. Extract with water to remove $Na_2C_2O_4$ and Na_3PO_4. Test for PO_4^{3-} and $C_2O_4^{2-}$ as in Steps 4 and 6. The solid remaining is yellow if $BaCrO_4$ is present. Add acid to the solid; if $CaCO_3$ is present, gas is evolved.

CHLORIDE GROUP ANIONS

101. a. $2\ AgCl(s) + S^{2-} \longrightarrow Ag_2S(s) + 2\ Cl^-$
     b. $Ag^+ + SCN^- \longrightarrow AgSCN(s)$
     c. $FeS(s) + 2\ H^+ \longrightarrow Fe^{2+} + H_2S(g)$

102. a. $2\ AgBr(s) + Zn(s) \longrightarrow Zn^{2+} + 2\ Ag(s) + 2\ Br^-$
     b. $2\ NO_3^- + 6\ I^- + 8\ H^+ \longrightarrow 2\ NO(g) + 3\ I_2 + 4\ H_2O$
     c. $MnO_2(s) + 4\ H^+ + 2\ Br^- \longrightarrow Mn^{2+} + 2\ H_2O + Br_2$
     d. $S_2O_8^{2-} + 2\ I^- \longrightarrow 2\ SO_4^{2-} + I_2$
     e. $Zn(s) + SCN^- + 3\ H^+ \longrightarrow Zn^{2+} + H_2S(g) + HCN(g)$

103. a. I^- (Br^- less readily).
     b. Br^-, I^-, S^{2-}.
     c. Br^-, I^-, S^{2-}, SCN^-.
     d. I^-.

104. a. Oxidize I^- to I_2 with $K_2S_2O_8$, acetic acid.
     b. Add H^+ to generate H_2S.
     c. Add $AgNO_3$ in acidic solution to precipitate AgI.
     d. Add Fe^{3+} to give a deep red color with SCN^-.

105. a. To avoid oxidizing Br^- and SCN^-.
     b. The I^- ion gives a faint red color with Fe^{3+}.
     c. This would precipitate Br^-, I^-, SCN^-, and other anions.

106. a. Step 1: There is a yellow   Step 3: Results are negative.
             color.   Step 5: There is a yellow color.
         Step 2: There is a violet   Step 6: There is a yellow color.
             color in the CCl_4 Step 7: A white precipitate is present.
             layer.   Step 8: Results are negative.

     b. Step 1: There is a yellow   Step 5: The solution is colorless.
             color.   Step 6: Results are negative.
         Step 2: There is a violet   Step 7: No precipitate is present.
             color.   Step 8: A black color appears.
         Step 3: There is a red color.

107. a. Add Fe^{3+} to give a red color with SCN^-. Heat the original sample with $K_2S_2O_8$ and H_2SO_4 to oxidize the Br^- to Br_2, which gives a yellow color with CCl_4. Treat the oxidized solution with $AgNO_3$ to precipitate $AgCl$.

   b. Add HCl to give a gas with CO_3^{2-}. If I_2 is liberated, both CrO_4^{2-}
      and I^- are present. Oxidize another sample with H_2O_2 to give a blue
      color if CrO_4^{2-} is present. Treat the sample with $AgNO_3$ in the
      presence of a strong acid to precipitate I^-.

108. Present: S^{2-}.
     Absent: Cl^-.
     Questionable: Br^-, I^-, SCN^-.

109. Present: SCN^-.
     Absent: S^{2-}, SO_3^{2-}.
     Questionable: CrO_4^{2-}, Br^-.

110. Extract KSCN with water; test for SCN^- with Fe^{3+}. Extract the remain-
     ing solid with HCl to bring CaC_2O_4 into solution; test for $C_2O_4^{2-}$ with
     $KMnO_4$. Treat the remaining solid with HNO_3 to bring CuS into solution;
     test for Cu^{2+} with NH_3. Any solid remaining after treatment with H_2O,
     HCl, and HNO_3 must be AgBr.

NITRATE GROUP ANIONS

101. a. $Fe^{2+} + NO_2^- + 2\,H^+ \longrightarrow Fe^{3+} + NO(g) + H_2O$
     b. $NO_2^- + NH_4^+ \longrightarrow N_2(g) + 2\,H_2O$
     c. $2\,MnO_4^- + 5\,NO_2^- + 6\,H^+ \longrightarrow 2\,Mn^{2+} + 5\,NO_3^- + 3\,H_2O$
     d. $Cr_2O_7^{2-} + 3\,NO_2^- + 8\,H^+ \longrightarrow 2\,Cr^{3+} + 3\,NO_3^- + 4\,H_2O$

102. b, d, e.

103. a. Add $AgNO_3$ to precipitate AgCl.
     b. Add HCl; bubbles form with CO_3^{2-}.
     c. Add dilute H_2SO_4 and $FeSO_4$ to test for NO_2^-.
     d. Add $BaCl_2$ to test for SO_4^{2-}.

104. a. Heat with MnO_2 in acid so-   d. Use an oxidizing agent, such as
        lution. NO_2^-.
     b. Add OH^-.   e. Dissolve in water; add $NaNO_2$ and
     c. Heat with NH_4^+. HNO_3 to precipitate AgCl.

105. a. Precipitate with $AgNO_3$.   c. Boil with H_2SO_4 and $K_2S_2O_8$.
     b. Heat with NH_4^+.   d. Heat with dilute acid.

106. a. Test 1: It may give off ClO_2. b. No reaction in any of the tests.
        Test 2: This converts I^- to I_2. c. No reaction.
        Tests 3, 4, 5: No reaction.   d. Step 1: No reaction.
                                             Step 2: Precipitation of AgCl.
                                             Step 3: No reaction.

107. Present: $C_2H_3O_2^-$, ClO_3^-.
     Absent: CO_3^{2-}, Cl^-.
     Questionable: NO_3^-.

108. Add acid and heat; if S^{2-} or $C_2H_3O_2^-$ is present, the characteristic odor
     is detectable. If S^{2-} is present, $C_2H_3O_2^-$ is difficult to detect. To test
     for SCN^-, add Fe^{3+} to the sample of unknown. To test for ClO_3^-, add
     $AgNO_3$ and $NaNO_2$ to form $AgCl$. To test for CrO_4^{2-}, add $BaCl_2$ to give
     a yellow precipitate if the ion is present.

CHAPTER 5

ALLOYS AND GENERAL SOLIDS

101. a. $Al(s) + 3 H^+ \longrightarrow Al^{3+} + \frac{3}{2} H_2(g)$
     $Zn(s) + 2 H^+ \longrightarrow Zn^{2+} + H_2(g)$
     b. $Sb(s) + 6 H^+ + 3 NO_3^- + 6 Cl^- \longrightarrow SbCl_6^{3-} + 3 NO_2(g) + 3 H_2O$
     c. $AgCl(s) + Cl^- \longrightarrow AgCl_2^-$
     d. $CaCl_2(s) + H_2SO_4(l) \longrightarrow CaSO_4(s) + 2 HCl(g)$
     e. $CO_3^{2-} + 2 H^+ \longrightarrow CO_2(g) + H_2O$
     f. $Zn^{2+} + CO_3^{2-} \longrightarrow ZnCO_3(s)$
     $ZnCO_3(s) + 4 OH^- \longrightarrow Zn(OH)_4^{2-} + CO_3^{2-}$
     $ZnCO_3(s) + 4 NH_3 \longrightarrow Zn(NH_3)_4^{2+} + CO_3^{2-}$

102. In water: KBr.
     In 6m HCl: Mg, CuO.
     In 6M HNO_3: Ag, CuS.

103. a. This is done to concentrate ions to get stronger tests.
     b. If these ions are present, Ba^{2+}, Ca^{2+}, and Mg^{2+} precipitate when the
        solution is made basic to test for Group III cations.
     c. Concentrated sulfuric acid reacts explosively with ClO_3^-.
     d. Some solid unknowns are insoluble in acid but dissolve with Na_2CO_3.
        Also, CO_3^{2-} removes certain cations that interfere with anion tests.

104. a. Step 1: It dissolves in aqua regia. A precipitate of $PbCl_2$ is obtained
        on dilution. The solution gives a positive test for Sn^{4+} in Group II
        analysis. Step 2: A positive test for Pb^{2+} is obtained.
     b. Step 1: It dissolves in HCl. No precipitate appears on dilution. The
        test for Group II cations is negative, but tests for Zn and Ni in Group
        III are positive.
     c. Step 1: It dissolves in HNO_3. A precipitate of BiOCl forms on dilution.
        The test for Cu^{2+} in Group II is positive. Step 2: The test is not
        positive. Step 3: Test for Bi^{3+} is positive.

105. a. Step 1: It is not completely water soluble. Step 2: A precipitate of $PbCl_2$ may form on dilution. Step 3: The tests for $C_2O_4^{2-}$ and PO_4^{3-} are negative. A precipitate of $PbCO_3$ or $CuCO_3$ forms on addition of CO_3^{2-}. (Not all Pb^{2+} precipitates in Step 2.) This precipitate is partially soluble in both NaOH and NH_3. Step 5: One may detect $NO_2(g)$. Step 6: Tests for SO_4^{2-} and NO_3^- are positive. Step 7: No additional anions are detected. Step 9: Identify $PbSO_4$ and $PbCl_2$.

    b. Step 1: It dissolves in HNO_3, perhaps giving a residue of sulfur. Step 2: A precipitate of AgCl forms on dilution. Step 3: Tests for PO_4^{3-} and $C_2O_4^{2-}$ are negative. A carbonate precipitate appears, which is insoluble in NH_3 and in NaOH. Step 5: One may be able to detect $NO_2(g)$; the liquid may turn yellow. Step 6: The test for NO_3^- is positive. Step 7: The test for SCN^- is not positive; since Fe^{3+} is already present, the solution is originally red. Step 8: This should precipitate Ag^+ and Fe^{3+}. The test for SCN^- is positive. Step 9: Identify AgCl and perhaps sulfur.

106. a. Attempt to dissolve it in water; if a precipitate is obtained, both salts are present. If it is water soluble, add $AgNO_3$ to test for $BaCl_2$. If there is no precipitate, add H_2SO_4 to a new solution to test for Pb^{2+}.

    b. Add water; test the water solution for SO_4^{2-}. Any residue must be $CaCO_3$.

    c. Add water; test the water solution for NO_3^-. Add H^+ to the residue and test for Fe^{3+}. Any residue remaining from the acid treatment must be AgI.

107. The white residue must be AgCl, indicating the presence of $AgNO_3$ and KCl. The absence of color in the solution eliminates the possibility of $CuSO_4$. The carbonate test is not proof of the presence of $Ca(NO_3)_2$ since $AgNO_3$ in excess would give a precipitate.

108. The fact that the mixture is colored indicates $FeCl_3$. The residue must not contain $PbSO_4$ since that is insoluble in acid. To test for Al, watch for the evolution of H_2 on treatment with HCl. To test for CaC_2O_4, add MnO_4^- to the acidic solution. Note whether it is decolorized.

PREPARATION OF REAGENTS

ACIDS AND BASES

In making solutions of acids and bases from commercial liquid reagents, add the indicated volume of reagent to distilled water to make the indicated volume of solution.

Acetic acid	17M	commercial glacial acetic acid
	6M	350 ml/liter solution
Hydrochloric acid	12M	commercial 37% HCl (conc HCl)
	6M	500 ml/liter solution (dilute HCl)
Nitric acid	17M	commercial 70% HNO_3 (conc HNO_3)
	6M	400 ml/liter solution (dilute HNO_3)
Sulfuric acid	18M	commercial 98% H_2SO_4 (conc H_2SO_4)
	6M	330 ml/liter solution (dilute H_2SO_4)
Ammonia	15M	commercial 30% NH_3 (conc NH_3)
	6M	400 ml/liter solution (dilute NH_3)
Barium hydroxide	saturated	$\sim$10 g $Ba(OH)_2 \cdot 8H_2O$/liter
		Shake well; let settle;
		decant clear solution
Sodium hydroxide	6M	240 g NaOH/liter solution

(6M NaOH solution is best prepared by dissolving the required amount of solid NaOH in a small amount of water. Then dilute to the proper volume.)

SALTS

Add the indicated amount of reagent to the given volume of distilled water (or specified solvent) and mix thoroughly.

Aluminum nitrate	0.1M	38 g Al(NO$_3$)$_3$·9H$_2$O/liter
Ammonium carbonate	1M	100 g (NH$_4$)$_2$CO$_3$/liter
Ammonium chloride	3M	160 g NH$_4$Cl/liter
	1M	54 g NH$_4$Cl/liter
	0.1M	6 g NH$_4$Cl/liter
Ammonium molybdate	0.5M	90 g (NH$_4$)$_6$Mo$_7$O$_{24}$·4H$_2$O/liter
Ammonium nitrate	0.1M	8 g NH$_4$NO$_3$/liter
Antimony chloride	0.1M	23 g SbCl$_3$/liter 3M HCl
Bismuth nitrate	0.1M	48 g Bi(NO$_3$)$_3$·5H$_2$O/liter 3M HNO$_3$
Cadmium nitrate	0.1M	31 g Cd(NO$_3$)$_2$·4H$_2$O/liter
Calcium chloride	1M	150 g CaCl$_2$·2H$_2$O/liter
	0.1M	15 g CaCl$_2$·2H$_2$O/liter
Chromium nitrate	0.1M	40 g Cr(NO$_3$)$_3$·9H$_2$O/liter
Cobalt chloride	0.1M	24 g CoCl$_2$·6H$_2$O/liter
Copper nitrate	0.1M	24 g Cu(NO$_3$)$_2$·3H$_2$O/liter
Iron(III) nitrate	0.1M	40 g Fe(NO$_3$)$_3$·9H$_2$O/liter
Lead nitrate	0.1M	33 g Pb(NO$_3$)$_2$/liter
Magnesium nitrate	0.1M	24 g Mg(NO$_3$)$_2$·6H$_2$O/liter
Manganese(II) chloride	0.1M	20 g MnCl$_2$·4H$_2$O/liter
Mercury(I) nitrate	0.1M	56 g HgNO$_3$·H$_2$O/liter
Mercury(II) chloride	0.1M	27 g HgCl$_2$/liter
Nickel nitrate	0.1M	29 g Ni(NO$_3$)$_2$·6H$_2$O/liter
Potassium bromide	0.1M	12 g KBr/liter
Potassium chlorate	0.1M	12 g KClO$_3$/liter
Potassium chloride	0.1M	7.5 g KCl/liter
Potassium chromate	1.0M	194 g K$_2$CrO$_4$/liter
	0.1M	19.4 g K$_2$CrO$_4$/liter
Potassium ferrocyanide	0.2M	84 g K$_4$Fe(CN)$_6$·3H$_2$O/liter
Potassium iodide	0.1M	17 g KI/liter
Potassium nitrate	0.1M	11 g KNO$_3$/liter
Potassium oxalate	1M	184 g K$_2$C$_2$O$_4$·H$_2$O/liter
	0.1M	18 g K$_2$C$_2$O$_4$·H$_2$O/liter
Potassium permanganate	0.02M	3.2 g KMnO$_4$/liter
Potassium thiocyanate	0.1M	10 g KSCN/liter
Silver nitrate	0.1M	17 g AgNO$_3$/liter
Sodium acetate	2M	280 g NaC$_2$H$_3$O$_2$·3H$_2$O/liter
	0.1M	14 g NaC$_2$H$_3$O$_2$·3H$_2$O/liter
Sodium carbonate	0.1M	11 g Na$_2$CO$_3$/liter
Sodium chloride	0.1M	5.8 g NaCl/liter
Sodium nitrite	0.1M	7 g NaNO$_2$/liter (not stable)
Sodium phosphate	0.1M	38 g Na$_3$PO$_4$·12H$_2$O/liter
Sodium monohydrogen phosphate	1M	142 g Na$_2$HPO$_4$/liter
Sodium sulfate	0.1M	14 g Na$_2$SO$_4$/liter
Sodium sulfite	0.1M	13 g Na$_2$SO$_3$/liter

Tin(II) chloride	0.1M	23 g $SnCl_2 \cdot 2H_2O$/liter
		Add 10 ml conc HCl/liter
		Add 10 g tin/bottle to stabilize
Zinc chloride	0.1M	14 g $ZnCl_2$/liter

SOLUTIONS OF ORGANIC SUBSTANCES

Aluminon	1 g ammonium aurintricarboxylate/liter of water
Dimethyl glyoxime, 1%	12 g DMG/liter of 95% ethanol
Magnesium reagent	0.1 g *p*-nitrobenzeneazoresorcinol in 1 liter 0.025M NaOH
Methyl violet	0.5 g/liter of water
Thioacetamide, 1.0M	75 g/liter of water

PURE SUBSTANCES AND MISCELLANEOUS

Aluminum wire	26 gauge, 1 inch lengths
Ammonium thiocyanate	NH_4SCN
Carbon tetrachloride	CCl_4
Copper wire	24 gauge, 3 inch lengths
Diethyl ether	$C_2H_5OC_2H_5$
Ethanol 95%	C_2H_5OH
Ferrous sulfate	$FeSO_4 \cdot 7H_2O$
Hydrogen peroxide, 3%	dilute 100 ml of stock 30% H_2O_2 to 1 liter (unstable)
Hydrogen sulfide, 0.1M	generate with 6MHCl and FeS and bubble for 10 minutes through water
Oxalic acid	$H_2C_2O_4$
Potassium nitrite	KNO_2
Potassium peroxydisulfate	$K_2S_2O_8$
Sodium bismuthate	$NaBiO_3$

INDEX